A Failed Trip to Reality
Tale of Inevitable
Model Breakdowns

*****

I0142426

By
Riyadh Alokaili

*****

RIYADH ALOKAILI

Dedication:

Forces of reality come in no better forms than these:

Nasser, Tarfa, Khuloud, Hesham, Yara, Yousef, & Sara

Gravity can take a back seat

Acknowledgment

Thank you, Mona Eendra, on Unsplash for the photo used as background for the cover page.

Introduction:

Concisely, the book is a product of a self-initiated effort to understand the world. It was not done to write this book, to submit a paper, to earn a degree, or because it is part of a job. The effort came purely from a strong desire to understand reality.

What better place other than the second paragraph of a book's introduction to confess that if this book contains a valuable contribution to humanity it did so undeservedly because it would have arrived there from the most unlikely place and person.

The author is not a physicist, mathematician, logician, philosopher, researcher, or a writer, but rather an avid fan of them all. This need not be stated of course, as those inadequacies will become painfully apparent in the writing.

This book started with one vision but departed from that original concept to something different. This evolution went through a few stages.

The first (unexpecting) stage is when the writing was a collection of random relatively unrelated ideas. Some ideas were sufficiently unsatisfying that the author couldn't let them go. Obsessing about them and a deeper look only uncovered more gaps and many of the ideas broke-down under careful examination. This sparked an interest to look for a satisfactory explanation, but none where fulfilling. This was a mental pre-occupation that gave birth to the next stage.

The second (determined) stage is when a light bulb moment was experienced and there was a genuine feeling that a solution to reality was discovered. This solution was called the "betweenness model" of reality. This model attempts to present an alternative solution to reality that may overcome some nagging inconsistencies, anomalies, problems, and paradoxes that specifically plagued the first stage mentioned above and our current understanding of reality in general. The model was thought to have the potential to be more than a gap filler but may have the potential to unify feuding theories on reality that don't seem to get along despite the best efforts of the brightest minds in the world trying to mediate and negotiate a peace treaty. Then difficulties were encountered in making the self-developed "betweenness model" coherent and it too broke-down similar to what has happened in the first stage. This also became a mental pre-occupation that gave birth to the next stage.

The third (sober) stage is the stage of examining why do reality models breakdown under stress. They all do! How may we resolve this conflict? Is there a way forward? Or is this repeating breakdown pattern inevitable? Insight from this stage allowed the author to present a defense of the second stage's "betweenness model" of reality that was abandoned before and keep it on life support.

The book will have a loose reverse chronologic theme, where the final sober third stage precedes the determined second stage followed by the unexpecting first original mindset stage.

Corrections, suggestions, and feedback are welcomed at (betweenness@yahoo.com)

Introduction:

68.

Keep believing:

The stage   **69.**
is set for the next great mind:

70.

Good start:

Quit projecting on   **71.**
to me what I am innocent of:

Untamable and   **72.**
playing by its own order or none:

73.

Truth martyrs are rare:

                                                            74.

Wish this one is wrong the most:

75.

Bold prediction:

76.

Bold prediction:

What is available is assumed to be what is desired, but   **77.**
what is desired is never available:

The best language   **78.**
is the one you speak, or is it?

Exposed nature leaves   **79.**
little room for the imagination:

Keep telling myself to stop looking for limi-   **80.**
tations, but they are everywhere:

Don't   **81.**

judge info by its factuality alone:

82.

Uncertainty principle:

83.

No terms, No winning:

If organisms were numbers,   84.
none would be a prime number:

85.

Wishful thinking:

86.

Kingpins:

Multiple realities   87.
don't exist but are undeniable:

It is unsettling, anytime a structure of a basic   88.
question may need to be revised:

Transla-   89.
tion does not mean identical:

90.

Here by myself:

                                                          91.

Disagreeable bunch (Part One):

92.

Hegel is wrong:

93.

Next gold rush candidate:

Consider it a "Repeat to me what   94.

you think I said?" type exercise:

Your mind is the ultimate "Photoshop" and abstracts are a popular   95.
brush stroke function, possibly unique to its human premium edition:

96.

The ultimate milestone:

Not all the pens and keyboards could make mental language whole,   97.
never to outgrow being an incomplete summary:

Language is a head-   98.
scratcher, but I'm glad we have it:

Not all fanta-   99.
sies have declared themselves:

We are all   100.
haunted by ghosts of cavemen past:

Our under-   101.
standing is partially satisfactory:

A first, with   102.
"fulfillment pending" status:

103.

Asking too much of tools:

   104.

Disagreeable bunch (Part Two):

105.

Fact-checking loopholes:

Being every-   106.
where can be a cloak of invisibility:

Why a state of not   107.
knowing that we don't know?

108.

Wrong us for the job:

Poster-   **109.**
ior probing digit vs thermostat:

110.

Off label use:

111.

Which one is real? Neither:

Asking too much of philosophy and perhaps   **112.**
too little of idea's body language:

Can physicists be trusted enough, to part ways with   **113.**
some well-established intuitions:

114.

Never is never right:

The un-conserved necessary follow a synthesis and   **115.**
is ruled by change and vanishing

                                                                               116.
Classification of understanding:

Forecasting a minimum   **117.**
fact cross-checking requirement

118.

Effect of fact cross-checking

                                                                               119.
Poor man's fact cross-checking

120.

Camouflaged metaphysics:

                                                                               121.
Convention boxed dynamics

122.

Legitimacy of questions

When   **123.**
parents are caught having sex

Meaning and ignorance   **124.**
from where no one expected!

125.

Omnipresent confounders

126.

Uniqueness paradox

Simulation twist ending (spoiler: it's us,   **127.**
not aliens, demons, or computers)

Irre-   **128.**
versibility of some knowledge

129.
A
root cause of all progress

130.

We can't be trusted.

Enslave-   **131.**
ment as measured by language

132.

A law in jeopardy

133.

Consciousness black holes

Prophecies will turn-   **134.**
out foolish, more often than not

135.

Identity's indefinability

136.

Convention Prison Break

Are conventions in-   **137.**
herited from producer to product?

CHAPTER 1: BACKGROUND AND HOUSEKEEPING TOPICS

1.   Orientation:

Chapter 1:   Background and Housekeeping Topics

Chapter 2:   The book will start with the author's third and final evolutionary (<u>sober</u>) stage of thought. An attempt will be made to explain why all reality models breakdown at the edges and under stress.

Chapter 3:   As a special case study: The "betweenness model" of reality will be presented (<u>second determined stage</u>)

Chapter 4:   The initial un-organized mindset before the "betweenness model" of reality (<u>firsts unexpecting stage</u>)

2.   Confessions and Disclosure

It might be appropriate to bring the reader's attention to the non-rigid and non-formal style of writing despite the utmost importance of the topic and the seriousness of the core claims. So, it is urged to be on the lookout for cheap attempts at humor, entertainment, and silliness sometimes, which hopefully are not taken too literally.

It is also relevant to state that this book is based on notes from a personal thought exercise. The mental treadmill that resulted in this book might suit some readers, but it might not be your thing if you are an elliptical person. Or in other words, this book is the equivalent of a person going out to the park and playing soccer, football, or basketball with regular folks, which is the preference of some people. But others might prefer to sit back and watch professional soccer, football, or basketball. This book is not for the latter high expectation group. Very few of us have a freakish mind that can dribble a ball off balance at full speed, an arm that can throw a spiraling football 50+ yards, or the height and athleticism to dunk from the free-throw line.

3. Why not present this opinion about reality to specialized establishments?

Since this writing has nothing to do with my livelihood and is largely a pastime activity I do in my spare time, there is no reason at all for it to be anything but pleasurable. Although I honestly try my best to be faithful to the topic, this work being wrong or unworthy is beside the point of doing it as long as it is a source of happiness.

Playing with thoughts is incredibly enjoyable. Ideas are a wonderful companion. The conceptual aspect of reasoning is a pleasure. Talking about it is fun. Writing about it comes in handy during downtimes. One thing missing I would prefer to have but is not available to me is easy access to philosophers or physicists to pick their brains.

Engagement ends at these points. Beyond these aspects of this mental exercise, it becomes homework like. Honestly, there is little interest to do an academic level review of the literature. That is the opposite of joy. There is also little interest to follow the regulations needed for a specialized establishment to consider looking at this work, such as writing letters to editors, filling forms and applications, and following other review and publication guidelines. That sounds like work. Hard pass!

Also, there is this faint voice that keeps saying, there are plenty of people doing it in the proper systematic way. Maybe a free-spirited method that is followed here might in some unlikely way give a perspective that would be less obvious or less possible by following the rigorous methods that establishments go by and expect.

4. Selected Claims & Conjectures:

About Truth, Understanding, & Knowledge:

1- *"All models of reality have a limited workable zone and inevitably breakdown when used beyond that zone"* or
"There will never be a theory of everything" or
"All theories are theories of less than everything"

For related segment see chapter 2.

2- *"All judgments are prepared from a prejudice sample"*

For related segment see heading numbers: 63

3- *"Unjustified Perceived purposefulness often occurs with spontaneity"*

For related segment see heading numbers: 64

4- *"Truth is never available to our consciousness directly and therefore escapes confirmation"*

For related segment see heading numbers: 65, 66, 83, 95, 97, 101, 103,109, 110, 116, 120, 121, 123, 124, 125, 129

5- *"Truth-seeking is an off-label mind activity"*

For related segment see heading numbers:110

6- *"Internally invisible inconsistencies are inevitable in all condition boxed dynamics, convention, or systems"* *"Conditions necessarily introduce inconsistencies"*

For related segment see heading numbers: 121

7- *"All wins are created by terms. Terms precede all wins"*

For related segment see heading numbers: 83

8- *"Logic is no exception to the point (1), as Logic has broken-down measurably"*

For related segment see chapter 2.

9-   "DNA seems to be a knowledge carrier that the definition of knowledge and theories of understanding don't pay attention to"

For related segment see heading numbers: 9783

About Humans:

10- "Nowhere do you exist where other stuff does not. Nothing is purely you, and everything (even your actions) is mostly or entirely not you"

For related segment see heading numbers: 84

About Language & Expressions:

11- "Language & Expression Tampering Hypothesis"" The mere use of expressions or language to describe meaning inevitably changes that meaning"

For related segment see heading numbers: 84124

About the Hypothesized "Betweenness Model" of Reality:

12- "Reality is fully accounted for through two fundamental irreducible qualities conservatism and betweenness, and through two fundamental entities non-in-betweens (of conserved entities) and in-betweens (or forces)"

For related segment see heading numbers: 18 through 27

13- "Betweenness, change, interaction, relationship, and force are fundamentally indistinguishable in reality"

For related segment see heading numbers: 18 through 27

14- "Possessing a betweenness or change qualities rules out being all conserved (unchanging) or forceless"

For related segment see heading numbers: 18 through 27

15- "Lack of betweenness removes change, rules-out force, leaving only conserveds to remain"

For related segment see heading numbers: 18 through 27

16- "Lack of a constant (conserved) quality is not possible. Pure betweenness is not possible"

For related segment see heading numbers: 18 through 27

CHAPTER 2: MODELS BREAKDOWN, THEY ALL DO

5.   What are the available models? Or vehicles of truth?

The shortlist:
- Intuition
- Philosophy
- Metaphysics
- Myths
- History
- Reason and logic
- Mathematics, language, expressions, and speech
- Experience
- Science and test/measurement requiring analysis

6.   Which ones have already broke-down and were deemed limited or unreliable?

The (B) list:
- Intuition
- Metaphysics
- Myths
- History
- Experience

Leaving us with the following potentially trustworthy (A) list:
- Philosophy
- Reason and logic
- Mathematics, language, expressions, and speech
- Science and test requiring analysis

7. Demonstration of model breakdown

Showing situations where these models' breakdown will be divided into two parts:

1. Narrow topic arguments. A group of random ideas that are collectively grouped in chapter 4 but are indexed below.

2. General argument: Analytical examination of a general generic model.

8. Narrow topic arguments

Chapter 4 describes and goes over specific examples when models and tools breakdown. A reference index is added here for the readers' convivence:

- Philosophy:
    See heading number: 62, 63, 65, 68, 71, 72, 77, 80, 83, 85, 90, 91, 92, 93, 95, 96, 97, 99, 101, 102, 103, 104, 106, 108, 109, 110, 111, 112, 114, 116, 120, 121, 124, 125, 126, 127, 129
- Reason and logic:
    See heading number: 62, 63, 65, 69, 71, 72, 77, 80, 83, 87, 90, 93, 95, 96, 99, 101, 102, 103, 106, 108, 109, 110, 111, 114, 116, 120, 121, 123, 124, 125, 126, 127
- Mathematics, language, expressions, and speech:
    See heading number: 62, 63, 71, 72, 75, 77, 78, 82, 83, 87, 88, 89, 90, 93, 95, 96, 97, 98, 99, 101, 102, 106, 108, 109, 110, 111, 114, 116, 120, 121, 123, 124, 125, 126, 127

- Science and test requiring analysis:
  >See heading number: 62, 63, 65, 70, 71, 72, 75, 77, 80, 83, 87, 88, 90, 93, 95, 96, 99, 101, 102, 103, 106, 108, 109, 110, 111, 114, 116, 120, 121, 123, 124, 125, 126, 127, 129
- (B) list:
  >See heading number: 62, 63, 64, 65, 66, 71, 73, 74, 77, 81, 83, 84, 86, 95, 96, 99, 100, 102, 103, 105, 106, 108, 109, 110, 111, 114, 116, 120, 121, 124, 125, 127, 129, 130, 131, 133, 135

One of the book's core claims is that they (models) all breakdown at the edge when taken outside their workable zone.

9. General Argument

The general case that all model's breakdown takes two approaches:

1- Reality accessibility
2- Convention inescapability

10. Reality accessibility

The target of all models is, to tell the truth. All of them claim to be telling the truth. They cannot all be right. So, what is the best way to tell which one is telling the truth?

The best way to tell if a model is telling the truth is to see if it agrees with reality. Put in a table form the analysis will look something like this:

	Model Compatible Result	Model Incompatible Result
Agrees with Reality	Justified Workability	False Un-Workability
Disagrees with Reality	False Workability	Justified Un-Workability

However, we have a problem here, because we do not have direct access to reality (explained in chapter 4), no one has, and no one will ever have such access. It is humanly impossible. Then, if direct access to reality is not available what can we use to evaluate any model's accuracy?

It depends on who you ask and what kind of thing you are evaluating. But since we are talking about reality in this book. The simple and best answer is measurement. Another good consideration is logic. Since we do not have direct access to reality, we rely on measurements to fill in its vacant place. It is not the same thing as reality, but it is the best that we can do at the moment. Does measurement overcome our inability to have direct access to reality sufficiently well?

Unfortunately, sometimes not, but it does a better job at limiting the subjectivity. So, this leaves us with models being judged by measurement or sometimes logic. Are there relevant concerns with measurements and logic that might impede its reliability?

Yes, measurements and logic are convention bound and so are the models that they are supposed to judge. Furthermore, there are shared omnipresent assumptions that are humanly impossible to control for, therefore, representing an ever-present confounder. Hence, we return to our original claim: models' breakdown, they all do! Because of this inescapable need to use some preferred or favored convention (e.g. measurement) to judge another convention (e.g. science) without controlling for all the confounders (e.g. time, space, language, consciousness).

Can it be argued differently? Yes, of course, see below.

11. Convention inescapability

Let us start with a generic model, any model. All models are based on conventions. For example, our visual apparatus needs eyes, light, and conducting nerves. Our auditory apparatus assumes ears, sounds, and oxygen. Science counts on lawfulness, language, and time.

(Side note before going forward: Is spacetime a safe assumption? Time stops in a black hole and space locality is undermined by quantum mechanics. So, the answer is maybe not. Just remember that nearly all models are counting on spacetime to be a safe reliable assumption. This is another ticking time bomb for model breakdowns waiting to happen)

It is impossible to isolate any model from the conventions it relies on and defines it.

So, we can organize our analysis with the following table of types of model breakdown:

	Model Compatible Result	Model Incompatible Result
Within Convention Limit	Workable Zone	Inexplicable Un-Workability
Outside Convention Limit	Rejected Workability	Explicable Un-Workability

We can see that there are three flavors of model breakdown (gray boxes):

1- Inexplicable Un-Workability: Here think of Newtonian physics unworkability with

Electromagnetism, where relativity stepped in to remedy the unworkability of Newtonian physics.

2- Rejected Workability: This may be a zone where models are unjustly treated

and might get blamed for the convention limitation, limitations the model is innocent of.

If we take logic as our convention. Here you can think

of quantum physics championing superimposition and undermining locality as violations of logic conventions. If you uphold the logic convention, then you have to reject or conceal superimposition and things being in more than one place at a time, hence masking parts of quantum physics.

3- Explicable Un-Workability: Not difficult to detect and explain.

Expectation remark: It needs to be stated that a breakdown in a model when it happens does not tell you what category it belongs to. It might or might not be obvious.

An additional example: Before we leave the above list. In (point 2 in the list) we used quantum mechanics as our model under interrogation. What if, we use logic as our model and see how it breaks down in the above example. If logic is our model under interrogation, it appears that its differences with quantum mechanics are of the zone (1) (Inexplicable un-workability zone) variety. In other words, if we take quantum mechanics (representing measurement) as our convention. You can think of logic objection of superimposition and it rejecting things possibly being in more than one place as a model incompatible objection, placing this breakdown in zone (1) (Inexplicable un-workability zone) in the table.

Brainstorming comment before going further: The author believes when it comes to theories dealing with reality, reaching zone (2) (Zone of Masked Workability) occurs when we leave a fundamental presupposition such as leaving the time domain. Conventions like logic breakdown at the boundary between time and the timeless, in the author's opinion.

A remark about breakdown category (1) & (2): Conventions are nothing more than models and models are nothing but conventions. So, designation (1) & (2) are a product of vantage point or frame of reference. They may be interchangeable.

The relevance of this interchangeability can be phrased in the question: What convention or model should we give more say in the matter? Two answers come to mind:

- Ideal unrealistic situation: Reality should always be our reference, but it is not and will never be directly available to us.
- Realistic practical situation: Measurement is

filling in for reality.

Measurements are all convention-based so a convention-free frame of reference is not humanly available. At a minimum, any measurement has to be within time and space. Reality is assumed but not obligated to be within time or space. Here Measurement is just another model or convention, but it is our best chance at estimating reality.

Hence, we return to our original claim: models' breakdown, they all do!

Key relevant point: For better or worst it seems that measurement as the preferred convention frame of reference is our best choice with the understanding that it comes with its baggage and limitations. We can also see and notice that logic can measurably be shown to breakdown.

Now that we went over the anatomy of model breakdowns and how all models have a breaking point. It is time to go over how to navigate our search of reality and truth and go over conflict resolution strategies.

12. Conflict resolution and beyond the breakdown.

How can we overcome breakdown?

Models need to have reasonable excuse strategies on two fronts:

1. How would the model in question fare with respect to measurement?

   • Measurements are necessarily within

their convention limits (e.g. time). Therefore, it follows that there might be situations where the measurements don't apply to the model of reality in question because:

- Specifically: The model's scope extends beyond those convention limits of measurement (e.g. spacetime)

- Generally: There is evidence that the reality we live in is sometimes space-less and timeless:

   a. Time stops in a black hole
   b. Locality and space are undermined by quantum physics
   c. Chapter 4 includes observations and arguments about the suspicious and patchy nature of time and space

2. How would the model in question fare with respect to logic?

   • We begin with our key observation above (We can also see that logic can <u>measurably</u> be shown to breakdown). So, if a model breaks down on logical grounds, it might be argued that the model might be non-sensical within the logic framework only but maybe not in reality. Maybe the model in question has valid extensions beyond logic's breakdown limit and logic might not be a valid frame of reference. Similar to what happened when quantum mechanics measurably violated logic.

These strategies do not prove a model of reality, but it does show that some models can be kept in the category of undifferentiated understanding (this category is explained elsewhere). Instead of completely abandoning them.

In summary:

Humans are of and live within "reality" or "the universe". On the other hand, human understanding is of and lives within a "model universe". We can replace the word model with what every best fits any particular human: observable, mathematical, measurable, logical, or mythical......

We can then summarize, express, or demonstrate these broad findings in the following way too:

> The Universe =/= The Model Universe at its fullest potential
>
> The Universe =/= The Observable Universe at its fullest potential
>
> The Universe =/= The Mathematical Universe at its fullest potential
>
> The Universe =/= The Measurable Universe at its fullest potential
>
> The Universe =/= The Logical Universe at its fullest potential
>
> The Universe =/= The Mythical Universe at its fullest potential
>
> The Measurable Universe =/= The Logical Universe

It follows that for us to understand "reality" or "the universe" this understanding must be in-part:

- Model-free
- Un-Observable
- Non-Mathematical

- Un-Measurable
- Illogical

Humans are incapable of that type of understanding. Therefore, we come back to saying model's breakdown, they all do!

CHAPTER 3: THE "BETWEENNESS MODEL" OF REALITY

13. Chapter explanation

This chapter is here as a special case where a virgin model of reality is conceived, developed, justified, stated, described in more detail, then applied to reality and life.

At some stage, the seams started to fall apart. Faced with the incoherence of the model, it seemed to have broken down but then a lifeline was constructed in the form of arguments in defense of it that kept it as a plausible solution to reality for the meantime.

The chapter can be explained in another way too. Let us take the OJ Simpson trial. It can be presented in two ways:

1- Short way: OJ Simpson was acquitted on two counts of murder, but maybe he did it?
2- Long way: Watch hours of the trial and live through it, moment by moment.

Using that same theme, this chapter can be presented in two ways:

1- Short way: The "betweenness model" of reality broke down, but maybe it did not?
2- Long way: Read the next 42 pages.

14. Belief to disbelief and back:

There are moments of unexpected insight or realization that force you to pause a moment or two to allow you to re-orient yourself and digest the surprising information. Those moments are rare, unsettling, thrilling, and often satisfying.

Several examples can be recalled. There was that moment when my father told me that a signature need not be artistic or beautiful. Before that, I spent hours trying to perfect a signature, thinking signatures were supposed to be aesthetically pleasing. Also, there was that other moment when I learned that black and white movies were filmed that way without color. Before that, I thought that movies were all originally filmed in color and become black and white as they age causing the colors to fade away.

But the moment I want to talk about is that moment I was told that solid objects like a hammer are mostly empty space. It was difficult to wrap my head around something solid and used to drive nails into stuff being mostly nothing (an atom is more than 99.99% empty space). The certainty of a hammer's compactness, density, gap-less-ness, and continuity had to begrudgingly be disbelieved, nevertheless.

That brings us to another certainty that had something similar happen to it. Certainty here is meant to represent the layman's naive (and mine) concept of certainty regarding science, not the scientific concept of certainty (to be fair scientist never admitted certainty but some of them are very good salespersons who know how to sell their theories in a way that implies confidence and certainty to untrained trusting regular us). The solidity and soundness of the scientific work of course got eroded in a way reminiscent of what befallen our fellow the hammer. Science appeared less certain and coherent than the layperson in me ever thought possible. The deeper I looked, the more unfinished many of the theories looked. To the degree that some looked increasingly indistinguishable from fantasy. The solid hammer moment has repeated itself. The once-revered unapproachable scientific body of work is human after all.

A big fan of the sciences and lover of scientists could not help but feel incredibly sad. But soon the blues were quickly replaced with a sense of opportunity. Maybe, just maybe, this is the opening that an average person with a wisdom crush might need to gather enough courage to present a contribution from outside the typical establishments, those establishments that claim dominion over the understanding of reality.

15. Informal Notable Cracks in Understanding (open source edition)

Some notable cracks in current understanding:

1- The mystery of consciousness
2- Quantum physics failing to describe macroscopic systems
3- Relativity failing to describe the sub-atomic systems
4- The contradiction of claiming that light is both particle and wave
5- Coming across an article that state the quantum physics cannot consistently describe the use of itself
6- Multiple articles that describe paradoxes that shake basic logical givens such as "opposite facts cannot both be true"
7- The double-slit experiment
8- The EPR paradox
9- Schrödinger's cat paradox
10- This one cracks me up: It is when believers hesitate a little to back something fully because it sounded weird even to the faithful. I am speaking about the non-locality that comes with quantum physics, where physicists seem not to want to bring it up or to not be able to come out and completely endorse it. Do not worry, this writing and the "betweenness model" of reality will take it upon themselves and do it for them.
11- Mind-Body Problem
12- See below for more personal additions

16. Informal Notable Cracks in Understanding (personal edition)

Here are some personal observations (This list is not populated with known or well-known critiques of understanding, but they are discussed in more detail in chapter 4):

1- The "Convention boxed dynamic" problem, undermines the credibility of human understanding.

2- The weirdness of space coming and keep coming from nowhere

3- The weirdness of time coming and going from nowhere

4- The weirdness of knowledge coming and keep coming from nowhere

5- Language, convention, abstract, mathematics, logic, algorithms, and other mental products are significantly limited. This further undermined the credibility of human understanding.

6- Assuming that nature is intrinsically uncertain, is problematic

7- Language, convention, and expression redundancies. The above segment mentioned one example of wave-particle duality. Another example that may not have been recognized before is the energy-mass duality, which seems problematic to the author as well.

8- Concepts such as infinity

9- Physical laws don't seem to account for the mental that well

10- Concern about the potential pernicious far-reaching role of the "expression tampering effect" and "language tampering effect"

11- Mental limitation and ceiling of understanding.

12- The unclear pairing of consciousness with reality

13- Scientism to the degree of dogma

So, where does this leave us? The once formidable does not look undisputable anymore. The stable does not appear so settled after all. So, let us walk you through the "conserved-force model" or "betweenness model" of reality.

17. Historical potentially relevant context

Perhaps these cracks and problems listed above are in some way similar to the problems that partially plagued pre-Newtonian concepts of force and motion. Thinkers then ran into errors because they did not recognize the non-obvious force of friction. This caused miscalculation in the understanding of nature and motion. Could there be further non-obvious forces still plaguing our understanding of nature? There must be!

18. Proof Outline

The proof of the "betweenness model" of reality will take on two forms:

1- An observational and analytical form. A Socratic approach in away. See headings number: 192021, 22, 23.
2- Linguistic form. See heading number: 24.

19. Questions to Self

Noticing cracks and problems resulted in so many questions. These were raised and forgotten, time and time again, too many times to count. However, some questions seemed to be recurrent and felt more important, these were:

1. Could some divisions of things be deeply meaningful? For instance, a division along the line of "change" and "no change" or constant. Matter, energy, momentum, angular momentum, and charge don't seem to change in the total amount. Then there are these things that vary such as space, time, language, abstracts, life, and consciousness.

2. Could the quality of some things coming and going from and to nothing be deeply meaningful? Space, time, consciousness, language, words, abstracts, universe, DNA, RNA, and ideas come and go from and to nothing.

If we clean that up more it seems that the presence or lack of "*change*" stands out as being deeply meaningful.

20. A Cartesian moment, maybe?

The moment that unchanged-change seemed like a deeply meaningful quality of reality, brought to mind an image of Descartes when he spotted that the act of thinking was deeply meaningful because it cannot be denied.

21. Preliminary analysis of what our understanding has done broadly

From the above haphazard self-inquiries, we came out with "change" as being important. Possibly fundamentally so.

Now let us move forward and try to be a little more organized by looking at human efforts to understand reality broadly while keeping the quality of "change" in mind and also trying to get a sense of how things that makeup reality are grouped. This birds-eye inspection seems to uncover a broad overarching pattern. Our discoveries seem to be divided into two groups depending on their definability:

1- The somewhat relatively clear and definable
2- The somewhat relatively unclear and ill-definable

Examples of group (1):

3. The standard available models have identified some

aspects of reality that a law of conservation can be applied to and allows us to describe and predict them like energy, matter, momentum, angular momentum, and charge.

4. The standard available models have identified a peculiar exclusive set of entities that apply action and reaction and called them forces.

The clarity and definability that we are interested in here are not the particular examples (because these at some level can be disputed, have shifted over the years and are less relevant to our discussion) but rather the quality of "conservatism" and the quality "forcefulness" is what interests us here.

Examples of group (2):

5. Then it gets messy for most things outside group (1). Examples here include space, time, abstracts, language, life, and consciousness.

The unclarity and ill-definability that we are interested in here is not the particular examples but rather the absence of a common quality except perhaps the lack of conservatism or that they all change.

Why are there things without a conservation quality or law called force and other ones labeled something else? Therefore, that separation of group 1 from group 2 seems unsatisfying. The second group seems as if it is a dump basket group that is filled with apparently inhomogeneous entities. Could we unify it under one umbrella with a reasonable characterizing quality? Could the separation of "changing" parts of reality into force vs non-force be truly unjustified? Is this separation adding an unnecessary complexity that we don't need and are we are better off simplifying by removing arbitrary separators?

Those feelings sparked a search for a resolution. The thought process was convoluted, intermittent, and unsystematic but maybe best illustrated and organized to some extent in the below table.

Group	Group 1		Group 2	
Sub-Group	Associated with a Conservation Property	Forces	Others (Non-Mental)	Others (Mental)
Examples	Matter/Energy/ Momentum/ Charge/Angular momentum	Gravity/ Strong Nuclear/Weak Nuclear/Electro-magnetism	Spacetime/ Life/ Genes	Abstracts/ Ideas/ Language/ Consciousness/ Knowledge
Constant in Total	Yes	No	No	No
Standalone	Yes	No	No	No
Changing in Total	No	Yes	Yes	Yes
Creatable	No	Yes	Yes	Yes
Destroyable	No	Yes	Yes	Yes
In-between	No	Yes	Yes	Yes
Interactive Entity	No	Yes	Yes	Yes
Direction	Mostly Scalar	Yes	Mostly Yes	Yes
How may it cause change	Becoming	Being	Being	Being
Symmetry analysis	Intimate Association	Mixed	Mixed	Mostly Not symmetric	Mostly Not symmetric
Pairing with reality	Unsure	Unsure	Unclear	Unclear
Mind-Body Problem	Unclear	Unclear	Unclear	Unclear
Simulation	Unclear	Unclear	Unclear	Unclear
Magnitude	Yes	Yes	Somewhat	Yes
Equilibrium	Not sure	Applicable	Not sure	Not sure
Dimensional analysis	Confounded?	Confounded?	Confounded?	Confounded?

Each shade of gray highlights similar answers in each row. Notice that there are two solid black dividers. The right one divides the two groups and the left one divides group 1 into two sub-groups.

Let us next take a look at these dividers as to how they relate to the shades of gray. The left black column more often than not has different shades of gray on each side. The right column more often than not has similar shades of gray on both sides.

22. What may the grays tell us?

It makes it feel as if the left divider has a deeper meaning and ought to command more confidence. This is restating what we started with when we spotted that the quality of unchanged-change is deeply and fundamentally meaningful.

While the right divider might not be meaningful and might not deserve much confidence. The right divider might be illusory. Then we must ask ourselves: Is the separation of changing parts of reality to forces and non-forces justified?

What if we just removed the right divider like how the aether was removed from the discussion after relativity made it unnecessary?

What if we acknowledged that the right three columns have more in common than previously realized (as their shade of gray might imply) and might be related? Could they be unified somehow? Similar, to what relativity did in merging energy and mass.

These observations felt that they are in search of a new model of reality to account for them. This writing will humbly present one for consideration.

The "betweenness model" or "conserved-force model" of reality is an attempt to share a new way of viewing the world while trying to be sensitive to the above-described observations.

Limitations and final thoughts: The table seems revealing and insightful in the broad sense to the author. The conclusions that followed are not without some level of reason. However, this kind of evidence presentation and meaning extraction does not of course rise to the rigorous scientific, mathematical, algorithmic, algebraic, geometric, simulation backed, or experimental grade level of proof. But it still carries enough level of coherence and flow to be worthy of describing and sharing. The rigorous methods of proof are beyond the ability of the author, who hopes that there will be enough traction and appeal to the "betweenness model" that more capable minds might apply their tools and skills to disprove or validate it.

23. What important qualities with deeper meaning does the table suggest?

The following qualities stand out on a quick inspection:

1- Being conserved.
2- Changeability
3- Being Creatable
4- Being Destroyable
5- Betweenness
6- Independence and being a standalone entity
7- Being an Interactive Entity
8- Directionality

Some of these qualities appear derivative and some flow from other qualities. However, two qualities seem to stand out and appear essential qualities that the other qualities are related to or can be derived from:

A. Being Conserved/Un-Changeability
- This covers points 1, 2, 3, 4, 6
B. Betweenness
- This covers points 5, 7, 8

Here, we can see a subtle implication in this subgrouping. If (A) has a quality of Un-Change, then (B) is intimately associated with Change. It follows that betweenness and change are strongly related and might be two faces of the same coin. The author cannot find a clear decisive distinction where betweenness can ever be devoid of change and vice versa. Therefore, will be considered interchangeable.

These two qualities (conserved and betweenness) can be used to characterize the groups in the above table. The left-most column will become the group that is best characterized by the first quality of conservatism (unchangeability). The groups in the three right columns can be restructured into one group unified by the second character of (betweenness or change-ability). From now onwards this expanded group will be label "forces" that include the three right columns because they are considered forces nominally and literally according to the "be-tweenness model" or "conserved-force model" of reality.

It is, therefore, a claim of the "betweenness model" of reality that the distinction of changing entities into forces and non-forces is arbitrary. This division makes us ignore or overlook their relatedness through the qualities of betweenness and change.

24. Linguistic proof of the "betweenness model" of reality

There is a suggestion that the "betweenness model" of reality is hiding in the current standard definition of "force". Looking closely at the definition of force we can see that it is a description of force that might need to be revised.

Below is a step by step outline on how to deconstruct the current definition to bring out the "betweenness model" of reality from its hiding place. The inclusion of the phrase "change" (or physical effect) in the structure of the definition might be fundamentally unnecessary:

1. The current definition of force is:
    "A force is any <u>interaction</u> that, when unopposed, will <u>change</u> the motion of an object"

2. The two keywords (interaction and change) are inter-related and non-independent

3. Or we can say, the word "interaction" implies change:
   An interaction without change is ~~difficult (scratch that)~~ impossible to envision
   &
   It is hard to believe that the definition meant a Platonic type relationship

4. Therefore, it follows that there is a meaning overlap of the two words, in a subtle inter-dependent or self-referencing way

5. Cleaning up the definition without the suspected inter-dependent or self-referral part gives us the following definition:

   "a force is any <u>interaction</u>"

6. Great denial is required to not see an "interaction" as being "between" in essence.

7. This gives us partial justification to perform a word switch in the definition that gives us the following definition:

   "a force is any <u>in-between</u>"

8. That definition sounds similar to what the "betweenness model" of reality is trying to say.

25. Remarks on the proofs

Admittedly the two analytical and linguistic proofs of the "betweenness model" of reality are not airtight for it to fulfill the knowledge definition of "justified true belief". It is also admitted that reasonable doubt and objections can be argued on multiple accounts.

So, where does this leave us? It leaves us in a position to

explain why bother to describe it in the first place. It is described because the author believes that the claims are novel and by sharing it, it might cause some people to be aware of something they never noticed before and maybe see the profound potential of the betweenness point of view if it were true.

Below (or next) is a description of how the author envisions such a model might look like and how it might be applied to life and to answer some important questions.

26. Building block elements of the "Betweenness model" of reality. What are the types of real?

In keeping with the same theme, we can move forward and hypothesis that there are only two kinds of things in reality:

1- "Non-in-between" things. Can also be called Conserved entities
2- "In-between" things. Can also be called Un-conserved or Forces

For lack of a better term "betweenness" is used to describe a relationship that might or might not be dimensional or spatial in quality. Betweenness cannot start or end in a force it must start and end in a non-in-between or conserved entity.

The above example entities in the table (Third Row: energy, mass, momentum, charge, gravity, spacetime, abstract, language, ... etc.) are man-made concepts that are useful as a starting point of the discussion about the "betweenness model", but are limited for us to continue using them forward as representatives of the conserved for later discussions because they are particular and will fail to convey the essence of the "betweenness model" of reality. Therefore, there will be a need to introduce a more general "betweenness model" friendly and compatible terms and definitions.

Specifically, we can call these kinds of elemental things conserved entities (or non-in-betweens) and forces (or in-betweens).

Conserved entities are standalone unchanging entities that are defined by themselves regardless of others.

Forces are a non-standalone entity that exist in any in-between. Forces are defined by others and never by themselves alone.

The below illustration paints four scenarios for us to imagine:

1. In scenario (1) there is nothing.
2. In scenario (2) there is a solitary conserved entity. This is consistent with the definition in that it is a standalone non-in-between entity, that can be defined hence we illustrated it with a dot.
3. In scenario (3), there are two distinct conserved entities. By definition, a force must exist in any in-between, hence earning a designation with a dotted gray line on the illustration. A force can never exist in scenario (1).
4. That brings us to one more scenario in that a force may extend to and from a single conserved entity, see scenario (4)

To be clear the term "conserved" does not mean energy, mass, momentum, or charge. It describes elemental building blocks of reality that manifest in our observable reality in several ways including energy, mass, momentum, and charge.

Also, notice that the definition of force here does not require a humanly measurable action or reaction (e.g. gravity is measured in Newtons/Kg), it is assumed that these forces all have action and reaction capability, and states of equilibrium are possible, but some are below the threshold of human detectability or definability.

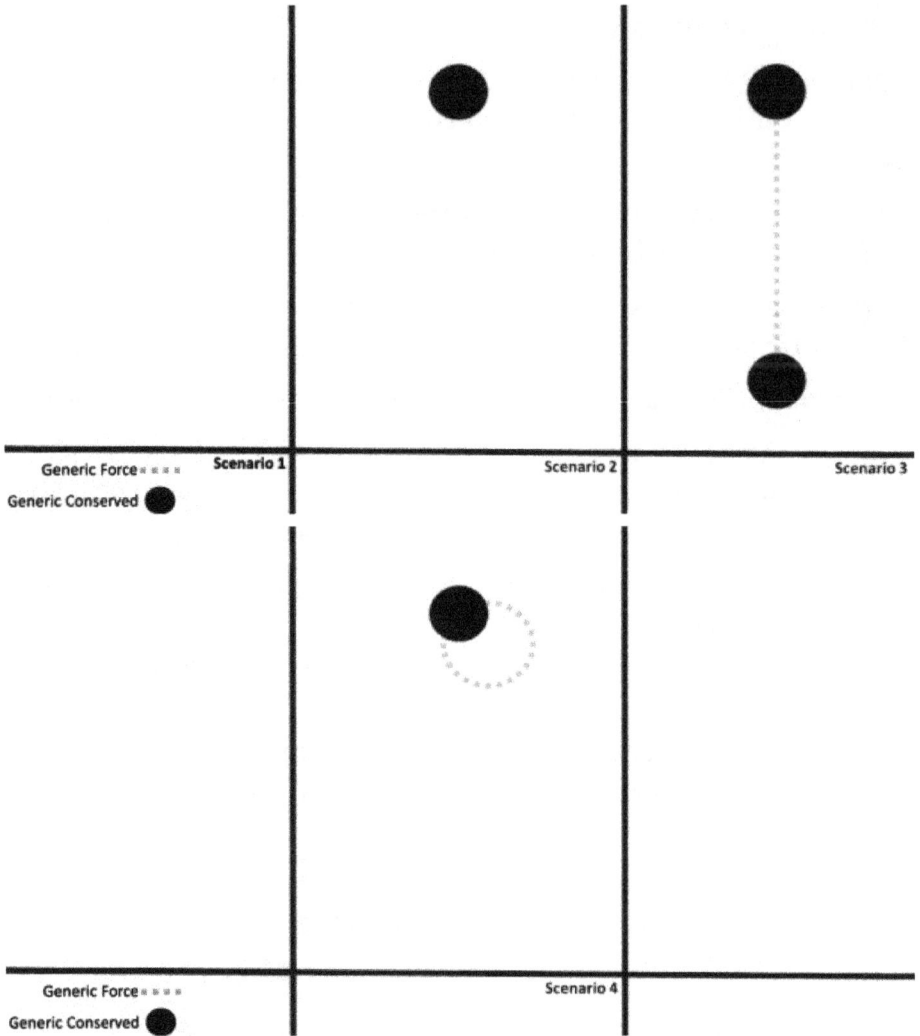

Generic Force ▪ ▪ ▪ ▪   **Scenario 1**   Scenario 2   Scenario 3
Generic Conserved ⬤

Generic Force ▪ ▪ ▪ ▪   Scenario 4
Generic Conserved ⬤

27. The "Betweenness Model"
of Reality can be stated as:

"Reality is fully accounted for through two fundamental irreducible qualities conservatism and betweenness, and through two fundamental entities non-in-betweens (of conserved entities) and in-betweens (or forces)"

***

"Betweenness, change, interaction, relationship, and force are fundamentally indistinguishable in reality"

***

"Possessing a betweenness or change qualities rules out being all conserved (unchanging) or forceless"

***

"Lack of betweenness removes change, rules-out force, leaving only conserveds to remain"

***

"Lack of a constant (conserved) quality is not possible. Pure betweenness is not possible"

***

The three key descriptors or words are real, change, between.

28. Implied in the "betweenness model" of reality

One noteworthy implication of the model that might be lost in the discussion is that the way we identify forces through their quality of causing a "physical effect" (or change) might be too high of a bar to not result in the exclusion of forces without a "physical effect" or those with a "physical effect" that is not observable.

"Betweenness" and "changing" are qualities of force that are independent of "physical effect" and therefore more inclusive.

As an example of exclusivity (maybe excessive exclusivity) of the current definition of force is that it sometimes kicks out long-standing established forces and revoke their force club membership. As illustrated by gravity being stripped of its force status and demoted to being a fictitious or a pseudo-force (it even goes further than this, in some points of view, all fundamental forces are fictitious). Gravity's honor is at stake.

The "betweenness model" of reality might be a way to right an unfortunate wrong. Join me to put an end to this and other unjust denials of rights. Gravity's cry's will be heard, we should never forget its last words "I can't ~~breathe~~ pull". We should hold whoever chocked gravity's ~~last breathe~~ pull away accountable for their actions. Support the movement of "gravity's ~~lives~~ betweenness matters".

It is the position of the "betweenness model" of reality that any entity with the quality of "betweenness" or "change" can be confidently assumed to be applying a force even if it cannot be detected.

Practical limitations of insisting on the physical effect requirement:

1. A force with a too-small signal may be unnoticeable
2. A background noise that is too large may mask a force from being noticed
3. Forces at equilibrium might be invisible
4. In addition to the practical limitations, there is also the problematic addition of "change" (or physical effect) in the definition of force. See heading number: 24.

It is also the implication of the "betweenness model" of reality that:

Interaction = Physical Effect = Change = Betweenness = Force

But some qualities are easier than others to notice and understand. And instinct might be behind not seeing the interchangeability similar to instinct being behind us not noticing earth's roundness.

If the scientific community adopts a "betweenness" metric it might play a complementary or substitute role to the "physical effect" quality of force. The following table contrasts the trade-off balance associated with insisting on the need for physical effect compared to adopting a betweenness metric alone or in the mix:

	Specificity/Sensitivity Trade-Off	
Force Quality	Physical effect	Betweenness
Specificity to Forces	More	Less
Sensitivity to Forces	Less	More

This understanding should lead to force subdivision into two groups:

1. Loud force: Has a definable change causing a physical effect
2. Calm force: Has no definable change causing a physical effect

There is a further subdivision of Loud-Forces as follows:

	Detectable Betweenness	Indetectable Betweenness
Detectable Change	Loud-Close Force	Loud-Distant Force
Indetectable Change	Calm Force	*Conserved*

When you get fond of a new idea so much you are setting yourself up for a "Baader-Meinhof phenomenon" or a "frequency illusion". The world looks different to me now. It seems that the world subconsciously was in waiting for a betweenness model long before this book. It seems as if the world anticipated the betweenness model. Take this attention-grabbing quote by Oliver Wendell Holmes for example:

> *"Between two groups of people who want to*
> *make inconsistent kinds of worlds,*
>
> *I see no remedy but force"*

Notice the words "between" and "force" have intertwining meanings and some level of synonymity. The world is looking increasingly "betweenistic".

Is this a mind game illusion? Not sure, but maybe. But very sure of one thing, the pleasantness of this world view and all the thinking it generated.

29. Types of conserved

The conserved elements are likely unknowable in the traditional scientific sense because they exist outside of time (time is considered a force by the "betweenness model" of reality, therefore, conserved entities cannot exist within a force or in other words cannot exist in time, in other words, conserved are timeless) in all likelihood and therefore we will suffice by saying they are standalone unchanging entities with no in-betweenness quality.

30. Types of forces

There are many ways these can be sub-divided. One subdivision in the above segment talked about loud-force and calm-force.

Forces or the in-betweens can also be subdivided into:

1- Mental forces:
   a. Where the force bridges do not extend outside of the brain. Examples include: experiencing ideas, experiencing memes, experiencing abstracts, experiencing conventions, experiencing language, experiencing mathematics, experiencing logic, experiencing algorithms, experiencing algebra, experiencing color, experiencing taste, experiencing emotions, and consciousness.
   b. For now, this is an example of a calm-force, until a quantification method is discovered that overcomes our current inability to measure it. If that is overcome, it will become a loud-force.

2-   Non-Mental forces: where the force bridges can exist inside or outside of the brain. Examples include gravity, electromagnetism, weak nuclear force, strong nuclear force, sound, time, and space. It also includes: expressing ideas, expressing memes, expressing abstracts, expressing conventions, expressing language, expressing mathematics, expressing logic, expressing algorithms, expressing algebra, expressing color, expressing taste, and expressing emotions.

31. What is not Non-Uniform?

The universe is not smooth. In the setting of our hypothesized real in the "betweenness model". The real is conserved, timeless, and immortal. Then, what does non-uniformity of the universe mean? It is a strong indication of the existence of more than one type of conserved. Because it is difficult to imagine a single timeless and immortal conserved elemental entity not being uniform. Where does the diversity and complexity come from? It seems to have been cultivated by the existence of two or more elemental conserved things.

Without a way to prove it, it is suspected that a tip of an iceberg metaphor is fitting when it comes to the number of conserved entities in reality. Everything we know, everything we can think of and everything we experience combined are confined to the tip of the iceberg that pocks above the surface of the observably possible loud-forces while it sits on top of a "below the surface" much larger group of conserved entities and calm-forces that are not observable. Complexity probably comes from   this.

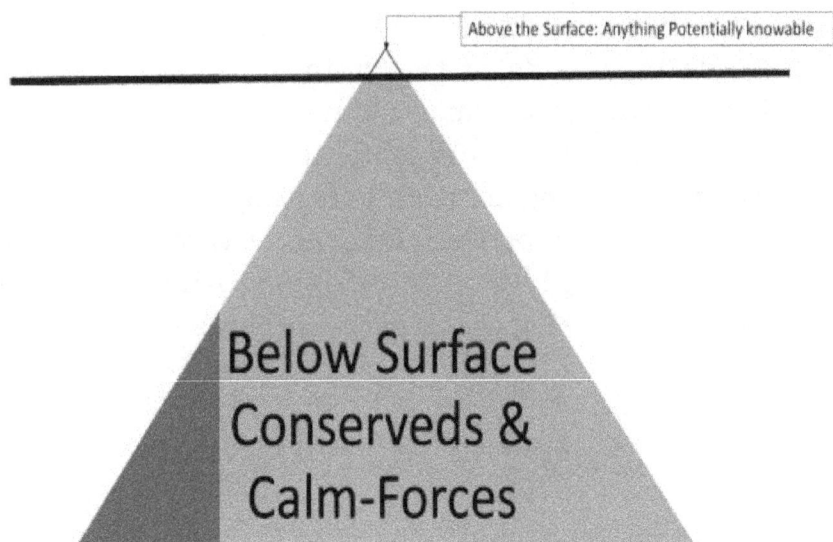

Above the Surface: Anything Potentially knowable

Below Surface
Conserveds &
Calm-Forces

32. Brainstorming about creation dynamics

There is no creation (at least within time) in the conserved domain (sense change is unconservative).

Creation pertains to the un-conserved only by definition (since it involves a change). It can also be stated that all creation requires a start and therefore, is un-conserved.

Within the creation or un-conserved or changing domain we can try next to subdivide this domain into two subtypes based on definability:

1. Definable unconserved
   - Convention (meant in this writing as mental modeling, organization and systemization products like language, logic, mathematics, etc....) does not pertain to the undefinable part of the un-conserved, by definition.
2. Indefinable unconserved

Therefore, there are undefinable creations and undefinable creative   processes.

Mental Ceiling to Comprehension → ← Physical Barrier to Comprehension

Unconserved
(Partially Definable)

Conserved
(Completely Indefinable)

Definable Creation | Indefinable Creation

No Creation

Definable Energy/Momentum?

Comprehensible Incomprehensible
(Convention Definable) (Convention Indefinable)

Convention Horizon Or The Mental Edge (Mentally inaccessible or blind spot to the Right)

Reality Horizon Or The Edge of Time (No Time or Creation to the Right)

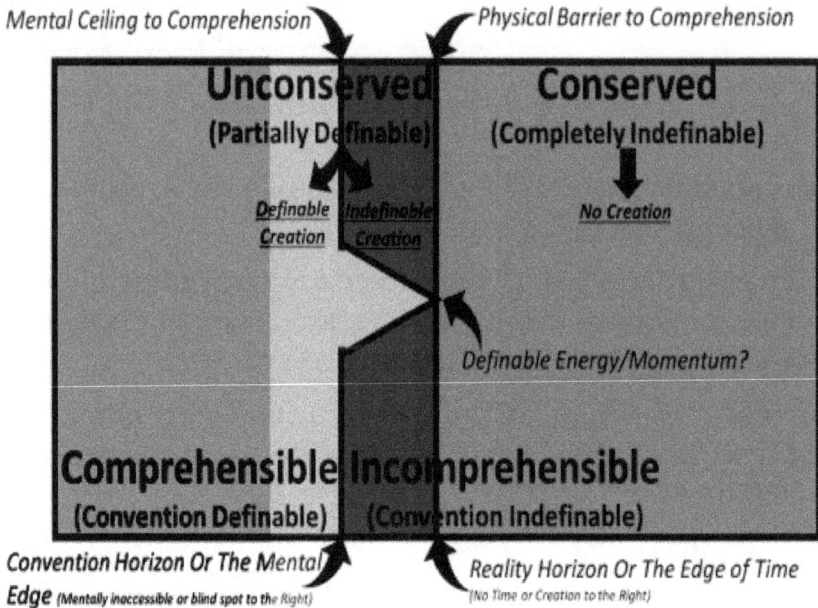

A product cannot fully understand its creator and therefore cannot fully understand itself and therefore it cannot create itself (at least within time) without a shortcut. We can put this differently, in two example form:

1- We are made by reality. We cannot understand reality in certain terms. Therefore, we will never be able to recreate ourselves from scratch.

2- We are made by reality. Reality is made of conserved elements. We can never make a conserved element. Therefore, we will never be able to recreate ourselves from scratch.

Therefore, for the purposes of this writing, we will reframe from using creation, since it pertains to the unconserved forces, and does not pertain to the conserved building block elements. Instead of creation, we will be using repurposing (or recycling).

33. Speculation about the source of repurposing

Warning: This segments organization might be confusing, so to help with orientation it will be divided into three parts:

A. An initial attempt at explaining the source of repurposing (very dense) (read at your own risk)

B. Model breakdown summary statement about an apparent "dead end"

C. An attempt at mending the model and conflict resolution

A.   An initial attempt at explaining the source of repurposing

As just stated, for the purposes of this writing we will try to reframe from using creation and use repurposing (or recycling) instead.

What follows is highly speculative since the conserved are timeless and not humanly comprehensible.

That said, let us imagine how the elemental fabric of reality (according to the betweenness model) might be classified according to an exist-repurposing ladder:

1- Un-Repurposed things: are defined as the unorganized unarranged conserved elements (we can call these mother elements).

Some arbitrary qualities of mothers:

i. Un-changing

ii. Stand alone.

iii. Not an interactive entity. Does not represent a relationship.

iv. Indivisible

v. Solitary

vi. Pure

vii. Stationary might be the unconserved

state for some

    viii. Timeless
    ix. Spaceless
    x. Forceless
    xi. Gravity-less
    xii. No betweenness quality
    xiii. Undetectable

2-   Conserved repurposed things: are defined as 1 or more purely conserved elements cultivating a compound conserved entity (we can call these daughter products). Here, the daughter product is a conserved arrangement of conserved elements. The best guess example for these "products" or "arrangements" is Total energy/matter, net charge, or/and net momentum. This is a guess because energy, matter, charge, and momentum have not met the unconditional knowledge requirements (described elsewhere) but will be assumed for practical purposes to be entities fully composed of conserved elements. The author suspects that the daughters' function as the connection between the mother un-repurposed with granddaughter un-conserved.

Some arbitrary qualities of daughters:

    i. Unchanging
    ii. Stand alone.
    iii. Not an interactive entity. Does not represent a relationship.
    iv. Indivisible
    v. Plural
    vi. Impure
    vii. Stationary might the unconserved state for some
    viii. Timeless

    ix. Spaceless

    x. Forceless

    xi. Gravityless

    xii. No betweenness quality

    xiii. Detectable as a category but not in individual units

3-   Un-conserved repurposed things: are defined as 1 or more conserved entities cultivating an un-conserved quality on top of the conserved constituent entities (we can call these differentiated granddaughter products). Here, the granddaughter product might be something like an un-conserved arrangement of conserved entities.

It is interesting to think of where does the un-conserved-ness comes from? It may be the arrangement that introduces the un-conserved-ness.

This category can be subdivided into two hypothetical categories based on the nature of the "arrangement"

    a. Conservative Granddaughter: The "arrangement" is carrying no energy, matter, and momentum. But it could be carrying another incomprehensible conserved player and therefore be yet another example of (1) or (2). Here, the above (3) definition is violated as all the elements of the definition are conserved with nowhere for the un-conserved-ness to come from.

    b. Liberal Granddaughter: The "arrangement" is clearly carrying no energy, matter, and momentum and it is also void of any hidden conserved component and therefore is a definition fulfilling example of (3).

Closest guess examples for granddaughter products are brains, DNA, RNA, biologic replication, cloning, budding, binary fission, and stars. Here there is, in general, no "conserved disconnect" or "conserved gap". In other words, no "father-product disconnect" or "father-product gap". In particular, there is no "matter or momentum disconnect" or "matter or momentum gap".

Something worth noticing is that they all have in common is that they are not mental. This repurposing/recycling process is non-mental.

Also, it might be reasonable to expect proportionality is found here more often than not.

It is never possible for a granddaughter product to be part of the cultivation of a daughter product, or in other words, granddaughter products are forever trapped to being granddaughters, disappearing, or evolving into an effigy product.

Some arbitrary qualities granddaughters:
    i. Stand alone.
    ii. Not an interactive entity. Does not represent a relationship.
    iii. Divisible
    iv. Plural
    v. Impure
    vi. No betweenness quality
    vii. Detectable individually

4- Interaction forces or entities or the non-repurposed: are defined as entities free of conserved elements (we can call these effigy products). Here, the effigy product is devoid of conserved elements. The best guess example is anything not containing energy/matter, charge, or/and having momentum. Examples include gravity, strong nuclear force, weak nuclear force, electromagnetism, spacetime, consciousness, abstracts, rules, and language. Here there is in general a "conserved disconnect" or "conserved gap". In other words, the "father-product disconnect" or " father-product gap". In particular, there is a "matter or momentum disconnect" or "matter or momentum gap".

It might be reasonable to suspect that disproportionality is found more often than not.

And also suspects that this creative process is mostly mental.

It is never possible for an effigy product to be part of the cultivation of a daughter or granddaughter product. Or in other words, effigy products are forever trapped to being effigies or disappearing.

Some arbitrary qualities of effigies:
  i. Not a standalone or cannot exist alone.
  ii. An in-between entity. It is an interactive entity. It represents a relationship. It acts as a bridge between two or more conserved entities or a connection between them. These are trans-entity interactive forces. For related segment see heading number: 35.
  iii. Divisibility is irrelevant
  iv. Numbering is irrelevant
  v. Purity is irrelevant.

Consciousness as we know it is nothing more than the sum of interaction forces.

Let us now look at a hypothetical machine-based consciousness. In the "betweenness model", biological consciousness is designated as an interaction entity. It is likely a sum of multiple individual interaction entities. The interaction is between two or more conserved entities (in the case of consciousness trillions of them is an understatement). In hypothetical machine-based consciousness the designers don't start with conserved entities and arrange them to create an interactive product similar to biologic consciousness, but instead begin with an interaction entity (code, program, language, mathematics, algorithms, binary, etc......), then the plan is to do a lateral move within the effigy product domain to produce artificial consciousness. There are a few problems with that based on the "betweenness model" or "conserved-force model":

1- It is taking a reverse engineering strategy where it "starts" with interactive entities instead of the biological way of producing consciousness which "ends" in interactive entities. AI consciousness will need to start with the un-conserved (code) to produce a special type of un-conserved (consciousness). Bypassing the conserved in the process seems questionable and probably impossible. If AI is to produce consciousness it need not create everything from scratch, but rather take a shortcut. Will that shortcut of being confined to the effigy products domain be enough? It looks doubtful. It seems that AI short cut must and needs to repurpose conserved entities on its way to consciousness. Similar to how parents' consciousness needs to retreat or re-anchor to DNA before there is going forward and repurposing of a child or offspring consciousness.

2- The well-established four fundamental interactive entities (loud- Force) probably play a role in consciousness and are part of the mix of elaborate interaction forces that gives us biologic consciousness. This is not accounted for in machines and AI hypothesized consciousness.

Mobility: The repurposing process seems to have repurpose traps where a product can only stay put in its cast or get downgraded in one direction from father to daughter, from daughter to granddaughter products. Effigy products are definitionally indefinitely bound to its cast, an inescapable part of these products' definition, identity, and fabric While other products have downward mobility through the casts.

Our life's complexity suggests that the number of conserved entities outnumbers all the un-conserved things we are aware of. To this point, this author is not sure of a single conserved entity, but energy and momentum appear to be the best examples so far, with a confirmation pending status.

This one-directional relationship between the products and the father always dictates that there are some absolute limits placed on the products but not on the father in absolute terms. Those limits necessarily mean a limitation in knowledge. This knowledge limitation necessarily prevents a product from understanding its real father. This places the product at a disadvantage and guarantees the father a privilege. This is what stands in the way and prevents a product from the ability to understand its father and the ability to make itself without re-anchoring in reality.

What if we hypothetically speaking, manufacture human DNA in a laboratory that was placed in a perfect environment to create a human with its consciousness? If conserved entities (including DNA) are in a suitable arrangement, the necessary interactive forces for consciousness will likely be met. It is not that different than a parent to child transfer of DNA. Again, there is re-anchoring and reconnection with matter (the manufactured DNA) which are materialistic substances that later cultivate a consciousness. No consciousness lateral transfers.

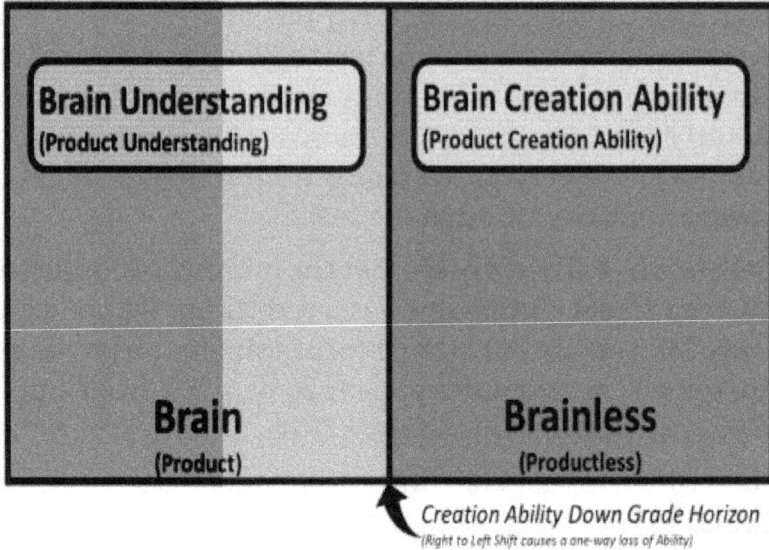

Brain Understanding (Product Understanding)	**Brain Creation Ability** (Product Creation Ability)
Brain (Product)	**Brainless** (Productless)

Creation Ability Down Grade Horizon
(Right to Left Shift causes a one-way loss of Ability)

B.　　　　　　Model breakdown summary

Dear reader, we started this segment (part A) having thought we figured out an answer to reality through the "betweenness model" of reality, then (part A) attempted to connect the "unchanging" conserved elemental building blocks to our "changing" existence. We have failed in developing a coherent connection or transition between the conserved and the un-conserved. If the elemental constituents of reality are unchanging and conserved, where does the change come from? Since the building blocks of reality are conserved unchanging elements, it remains unclear where does change, creation or repurposing comes from? This failure resembles to some extent the mind-body problem.

C.　　An attempt at mending the model breakdown
and conflict resolution

This defense is related to what we described in chapter (2)

under the title "Convention inescapability", see heading number: 11. It might be useful to revisit that segment.

Now let us go back to our main subject which is the "betweenness model" of reality. In parts (A) & (B) we attempted to explain the process of repurposing and then conceded that we ran into a dead-end where the betweenness model broke down. How can we overcome this breakdown?

The betweenness model needs to have a reasonable excuse on one or two fronts:

1. How would it fare with respect to measurement?
   - We can lean on measurements to some extent within time: The betweenness model of reality assumes that it bridges between two domains the timely and the timeless. Measurement on the other hand is mostly within time. No conclusive measurement is available in support of the betweenness model, but that does not mean that none exists. For example, we have some measurable interference observations that are predicted by the betweenness model that were collectively labeled the (expression tampering effect) (See heading number: 124).
   - We cannot lean on measurements outside time: The betweenness model of reality assumes that the conserved elements are timeless. Measurement on the other hand is mostly within time. Therefore, it follows that we cannot lean on measurement and there might be situations where measurement is inapplicable to the betweenness model of reality.

2. How would it fare with respect to logic?

- We stated earlier that (We can also see that logic can measurably be shown to breakdown). We suspect those measurable times that logic broke down was because those measurements were out-side spacetime. The "betweenness model" of reality might be non-sensical within the logic framework only. Maybe betweenness goes beyond the logics breakdown limit (outside spacetime), similar to what happened when quantum mechanics measurably violated logic (presumably when they went outside spacetime).

So, in summary in support of the betweenness model of reality, we leaned on measurement to some extent, we stressed that measurement and logic might not apply to all aspects of the model.

This does not prove the "betweenness model" of reality but it does show that it falls under the category of undifferentiated understanding (this category is explained elsewhere, see heading number: 116). I still have my fingers crossed that it might carry weight and adds to our human understanding.

How might we envision a sensible betweenness model that is not within the logic framework?

I don't know, we are not built to imagine well what we don't see in regular life. But three attempts to picture how that maybe can be are summarized in three comments that might be helpful:

1. The argument for "betweenness" might look like this: Only in our logic dominated human domain is something allowed to be change or change-

less. In reality, maybe, change and change-less are uncoupled and independent qualities of reality's elemental building blocks. Four scenarios of realities building block states might look like this:

- Unchanging only (sensical)
- Changing only (sensical)
- Both (nonsensical)
- Neither (nonsensical)

Here it is permitted to be conserved and unconserved at the same time, keeping the "betweenness model" of reality plausible. And overcoming the incoherent connection or transition between the conserved and the un-conserved. And solving the dilemma of where does the change come from an unchanging elemental entity?

2. Another way of trying to picture it is that betweenness as a quality appears within our human framework of understanding and experience. This quality collapses outside this framework where the elemental building blocks which truly have no betweenness quality but appear so to us from our vantage point. Imagine a world where all people have double vision all the time. Those people have always had two suns. Until the first person dared to use monocular vision instead of binocular vision. At that point, the two suns collapsed into one sun.

3. Another way of trying to make it more sensible is thru the below illustration that shows some things can have independent conserved and unconserved qualities:

- Conserved qualities:
  - Area

-    Color
  • Partially conserved or unconserved:
-    Number of Angles
-    Number of Limbs
  • Unconserved qualities:
-    Shape
-    Perimeter Length

So now that we have mounted a feeble defense of the "between-ness model", let us pretend it is a good model and continue to analyze and even dare to apply it to reality and life too.

34. When are lateral moves allowed?

It never occurs for the mothers and effigies. Consciousness is an effigy product and incapable of a lateral move to produce more consciousness. Program code is an effigy product and incapable of a lateral move to produce consciousness. I hate to have people angry at me, but I suspect that this news will not sit well with the technology-utopia cultists.

35. There are and must be more

than 4 fundamental interactive entities in the universe

There are well-established four fundamental interactive loud-forces. Some scientists are looking for a fifth one or even more. That remains to be seen. The four well-known ones are as follows:

1- Gravity
   a. Governed by the theory of general relativity
   b. Mediated by gravitons (hypothetical)
2- Electromagnetism
   a. Governed by the theory of quantum electro-dynamics
   b. Mediated by photons
3- Weak nuclear force
   a. Governed by the electroweak theory
   b. Mediated by W&Z bosons
4- Strong nuclear force
   a. Governed by the theory of quantum chromo-dynamics
   b. Mediated by gluons

The "betweenness model" or "conserved-force model" of reality suggests that the forces and interactive entities are much more numerous and possibly unfathomably so. Here are some suspected additional calm-forces or un-noticed interactive entities:

5- Space
   a. Maybe governed by a restructured theory of general relativity
   b. Mediated by TBD, maybe spaceons?
6- Time
   a. Maybe governed by a restructured theory of gen-

eral relativity
   b. Mediated by TBD, maybe timeons?
7- Non-Mental code or language of DNA and RNA
   a. Maybe governed by the theories of biology, cell function, and division
   b. Mediated by TBD, maybe nucleotides?
8- Mental interactive forces such as code, language, abstract, convention, rules.
   a. Governed by TBD, May be governed by a restructured neuroscience theories
   b. Mediated by TBD?
9- Consciousness is the sum of multiple interactive forces
   a. Governed by TBD, May be governed by a restructured neuroscience theories
   b. Mediated by TBD?
10- Life is the sum of multiple interactive forces
   c. Governed by TBD
   d. Mediated by TBD?
11- The Universe
12- An impossibly large number of other forces.

It is also suggested that the fundamental four interaction forces are composite forces and in due time will likely be further subdivided.

A metaphor that might be relevant is the clash of civilization or forces metaphor. Often seen as a clash between the West and East or Christian and Muslim civilizations or forces.

First of all, these imprecisely defined civilizations are accepted with little argument as factual. This might be reminiscent of the four fundamental forces.

Secondly and perhaps more importantly these divisions are hardly the only human divisions or forces. There are other civilizations if we allow ourselves to call them that. Geographically we can find clashing ones in South America, Africa, Asia, etc.

Historically there are clashing ones ancient and new. Intellectually, there are clashing civilizations of the scientific and non-scientific. Economically, there are clashing civilizations of the poor and rich or materialistic and non-materialistic. These later non-geographic examples might be reminiscent of the proposed unknown or unacknowledged forces (calm-force) in the above list (5-12). Realizing them is just waiting for a better or different point of view to see them as forces.

36. Betweenness and Conserved analysis: proposed new knowledge hunters

If the betweenness model caries validity, it suggests it might be beneficial to add a "betweenness analysis" and a "conserved analysis" as elementary techniques of analysis. It is hoped that it takes its place beside or instead of other elementary techniques such as "dimensional analysis", "symmetry analysis", "logic" or "algebra". These new proposed techniques seem worthy of consideration because they stand to be:

1. Fruitful (as exhibited by its explanatory potential)
2. Simple (it has limited assumptions, if not the least of any)
3. Wide scoping (it is not limited to physics but also touches on things such as language, ideas, and philosophy, where it has the potential to unify all of the existing physical, biological, mental, and other non-physical aspects of reality under one unifying theory)
4. Testable (as exhibited by the predictions and the model's clearly stated claims)
5. Admittedly and ironically, conserved analysis and betweenness analysis fails in being conservative as they are not expected to sit well with many established ideas and theories.

In brief, in the "betweenness analysis" everything that lacks a betweenness quality is a conserved entity, and anything that has a betweenness quality cannot be forceless.

In brief, in the "conserved analysis" everything that changes cannot be forceless, and things that do not change are a conserved entity. For example, thinking changes and therefore is forceful.

Paraphrasing the "Cogito, ergo sum" of Descartes we can say "I mostly think therefore I am forces mostly"

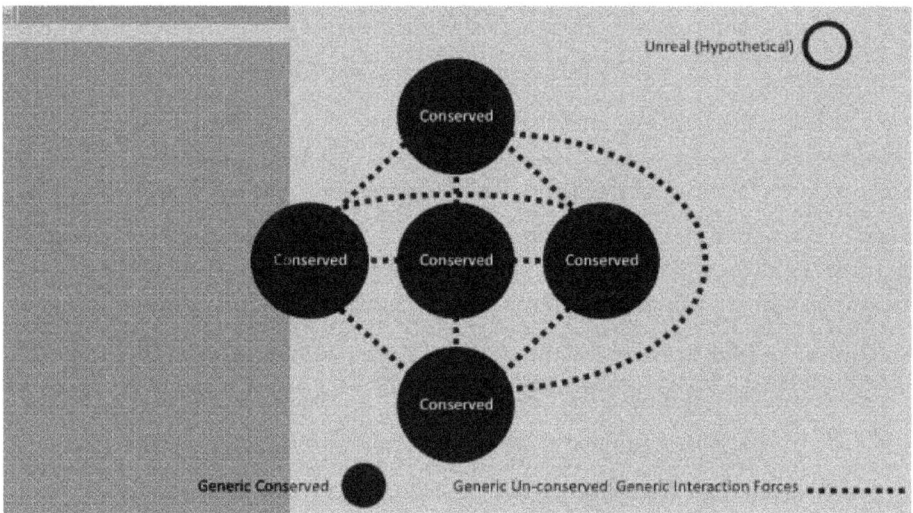

37. Conservation as an opportunity for understanding and discovery

This writing has outlined the difficulties in knowing what is real from the juxta-real and pseudo-real (see chapter 4). This writing has placed a lot of stock on the importance of being conserved (immortality and timelessness) as a quality of being real. Most things that we deal with do not appear conserved and therefore are suspect. If someone were to discover a way to make something conserved, immortal, and timeless, it is considered a confirmation of non-mental realness by the author. Relocating anything from the un-conserved category to the conserved category will likely be accompanied by great insight and understanding. For example, if someone were to show that time is conserved, then this author would have to revise this writing to something that aligns with this discovery where time has satisfied the non-mental realness requirement and its characterization as being doubtful and possibly just a mental concept become baseless because it is no longer un-conserved. The technique of conserved reassignment might benefit understanding and knowledge as much as the search for symmetry has benefited physics.

38. Short cuts to using and making things

Do we fully understand penicillin? but yet we use it. Did we make carbon? but yet we make carbon fiber.

A universal short cut that has never not been used is to repurpose already made stuff instead of making them from scratch. Another universal short cut that has never not been used is to utilize things without fully understanding them.

Humans have never made anything, nor have they understood everything. Making something literally requires full understanding and a full understanding of something requires making it. Which will never happen.

This is what one facet of the AI consciousness debate is hinged to. Is consciousness special and different from everything else we have ever made for it to need to start from scratch or can consciousness like everything else we have ever made, take a shortcut, and need not be made from nothing/scratch and need not be used or made with full understanding. From this, it seems that for AI consciousness to occur it only needs to be a re-purpose job not a matter of literal creation.

Should we stop using the words creation and making and start using repurposing or recycling instead? In pure terms, yes. The text however does use the phrases "create", "creation', and "creator" often. These are loose uses of the words and most are not literal.

39. A soft prediction from a highly speculative premature brainstorming session, part 1

In this writer's betweenness model, space may be a force or a composite of forces. What if space is a repulsive force that may in some way explain the perplexing expansion of the universe? What if it is the anti-gravity (and gravity is the anti-space) that happens to be weaker than gravity at "short" distances and stronger than gravity at "long" distances. Force and counter-force in a sense. Could this solve the mystery of us not finding dark matter and dark energy? Because they never existed, and they are not needed to explain the universe expansion if space and time are forces that satisfactorily explain that phenomena.

40. A soft prediction from a highly speculative premature brainstorming session, part 2

In this writing's model, time may be a force or a composite of forces. What if the time force counters the gravity force (anti-gravity) and the gravity force counters the time force (anti-time)? Could the immense gravity at a blackhole be because of the lack of the anti-gravity effect of time that leaves gravity unabated? What if we hypothesis a gravity-hole? A situation where there is a lack of gravity (anti-time), where the time force is unabated and become immensely strong. Could this be the equivalent of the hypothetical white-hole?

What if mass and spacetime are two faces of the same thing?

What if they are both daughter or granddaughter products? What if there is a subtle imperceptible loss of mass that is converted into space explaining the expansion of the universe.

41. Mind-Body problem solution

This writing's model of reality solves the mind-body problem. The brain is a composite of real conserved elements and all mental activities (e.g. consciousness, abstracts, dreams, language, etc.....) that do not contain conserved elements (e.g. mass, energy, momentum) is the interaction forces between the brain conserved elements. "Consciousness" is a composite force or interaction product of the "brain conserved elements" in the same way gravity is a force or interaction entity between masses such as the sun and earth.

It is understandable to view this model with surprise and judge it as being a weird concept. That view and judgment would not be the same if there was a good tool to measure consciousness or breakup consciousness into manageable component forces.

This solution to the mind-body problem also explains how abstracts affect the world around us. For example, the idea of a "currency" is a brain force that causes an action/reaction on brain granddaughter elements that might be sufficient to have external manifestations such as believing and valuing money. Language, abstract, mathematics, consciousness, and love are no longer mystical, transcendent, miraculous, supernatural, extra-dimensional, or magical things but are natural forces of reality.

It appears that the AI's approach to consciousness is not set-up for success as it is considering consciousness a product of calculations, this is inconsistent with the "betweenness model" or "conserved-force model" which views consciousness as a force that is a product of conserved elements interrelationship, not a product of calculations which is another force. A force cannot generate another force. To generate consciousness, you cannot start with forces, you need real non-in-between elements to interact with each other and have the type of complex relationship that weaves the composite sum force of consciousness. Consciousness is a property of reality, it is a force, and it, therefore, can apply and cause an effect on reality and change it.

This model has the explanatory power to eliminate the mind-body problem and replace it with another problem the conserved-force problem. In essence, the mind-body battlefront line has given way to a new battlefront line. This problem can be put in question form as "How does the conserved relate to the force?" We presented a solution earlier to this new conserved-force problem, see heading number: 33.

It is believed that progress has been made here, instead of just repackaging the same old problem or just accomplishing a lateral move. The new model is believed to be a step forward in our human understanding as it pushes the bottleneck of understanding from where it stagnated for centuries, to a more forward position and closer to the ultimate truth.

42. Outside of time

It is said that spacetime is the arena where everything happens. All the conserved entities are timeless and therefore outside time and therefore outside spacetime. But we know that energy is observable, therefore we have an observable entity that is dictated by a conservation law and therefore is at least in part timeless and therefore, outside time in part. The conserved may not be completely concealed after all and may have revealed itself or some of itself through energy. Energy might be our ambassador to the remaining conserved entities that are not currently observable and lie outside time. Never felt that our odds of getting to reality and truth are good, but energy gives us the smallest opportunity and the dimmest of light that not all hope is lost (momentum, charge, and angular momentum are other hopefuls).

43. Comprehension Blind Spots

Comprehension and understanding are qualities of consciousness.

Comprehension blind spots:

- A- The conserved (Certainly a black box):
  - 1- The conserved is a comprehension blind spot to the un-conserved including our consciousness (a comprehension handicap caused by fundamentally limiting consciousness conventions).
- B- The un-conserved (Possibly a black box):
  - 1- It either is convention definable then the answer is maybe?
  - 2- It is convention indefinable then yes, it is a comprehension blind spot to consciousness.

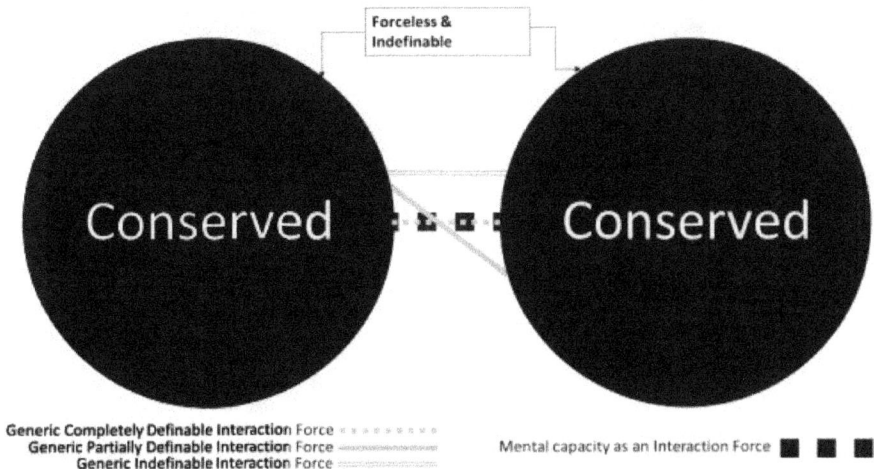

Forceless & Indefinable

Conserved · · · · Conserved

Generic Completely Definable Interaction Force · · · · · · · ·
Generic Partially Definable Interaction Force ~~~~~~~~
Generic Indefinable Interaction Force ════════

Mental capacity as an Interaction Force ■ ■ ■

- C- The unconscious (Subject to A & B):
  - a. If the unconscious is conserved, then yes.
  - b. If the unconscious is un-conserved:
    - i. It either is convention definable then the answer is maybe?

          ii. It is convention indefinable then yes, it is a comprehension blind spot to consciousness.

D- Consciousness (Subject to B but not A):

1- Consciousness is a force and has no conserved elements therefore A is inapplicable.

2- Then consciousness is un-conserved, and two situations exist:

          i. It either is convention definable then the answer is maybe?

          ii. It is convention indefinable then yes, it is a comprehension blind spot to consciousness.

Consciousness Free & Indefinable

Conserved · · · · Conserved

Consciousness Completely Definable Interaction Force
Consciousness Partially Definable Interaction Force
Consciousness Indefinable Interaction Force
Mental capacity as an Interaction Force ■ ■ ■

Schematically, there are three compartments with two dividers/separators to comprehension.

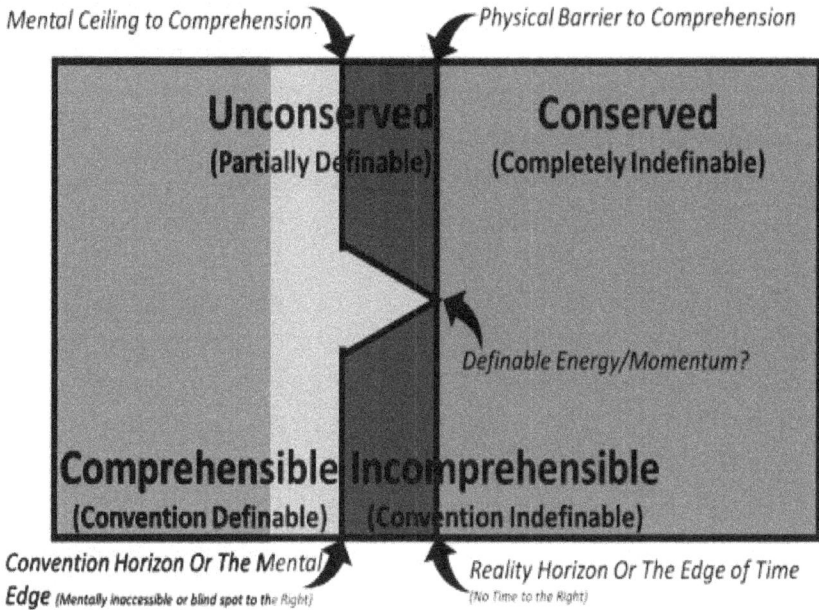

Mental Ceiling to Comprehension — *Physical Barrier to Comprehension*

Unconserved **Conserved**
(Partially Definable) (Completely Indefinable)

Definable Energy/Momentum?

Comprehensible Incomprehensible
(Convention Definable) (Convention Indefinable)

Convention Horizon Or The Mental Edge (Mentally inaccessible or blind spot to the Right) — *Reality Horizon Or The Edge of Time* (No Time to the Right)

The compartments are in shades of gray in the above illustration. The dividers are vertically oriented black separators, the first (right one) of which is the conserved/unconserved divide and the second (left one) is the convention definability/indefinability divide. Both are not precisely known and hard to determine. However, there are some important differences. It seems that we may be better at finding the conserved/unconserved divide compared to the convention definability/indefinability divide. This can be based on the characteristics listed elsewhere such as the character of timelessness and betweenness.

The above paragraph is a good segue to subdivide comprehension in a related but different way. Comprehension of course can be compartmentalized in many ways. The above compartmentalization just happens to be relevant, inspired, and compatible with this writing and the "betweenness model" or "conserved-force model" of reality that is being shared through this book. Another way to compartmentalization that is book and model friendly and inspired is to divide comprehension into:

1- Compartmentalized: This comes in three forms that match the shades of gray in the illustration above. The left-most compartment is where comprehension is possible in the above illustration.
2- Trans-Compartmental: Crosses the vertical separators. This is impossible (or we can say impossible within time).

How about consciousness? Consciousness is un-conserved (Therefore, state C below is inapplicable), and hence compartmentalized comprehension is feasible as long as it remains on the left side of the convention definability/indefinability divide (State B not State A). Trans-Compartmental understanding is impossible.

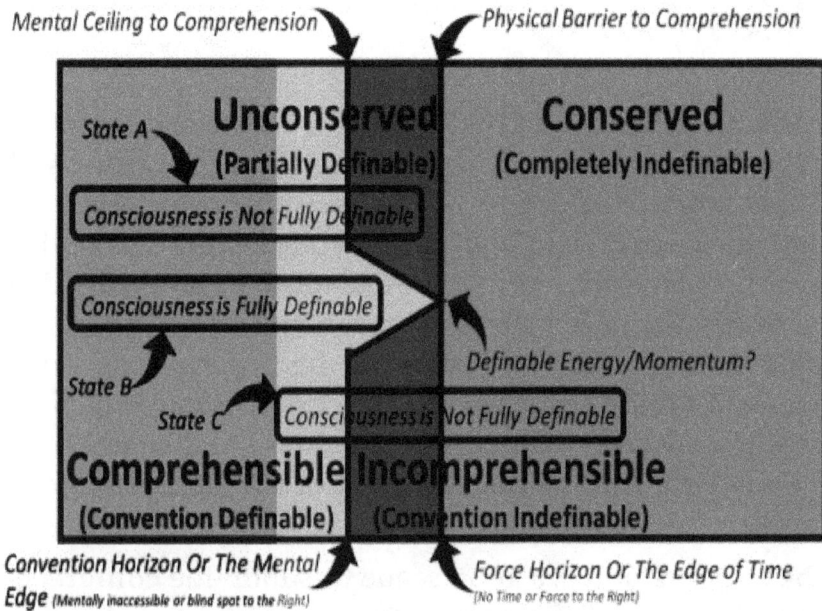

The above paragraph is yet another good segue to add another way to subdivide comprehension using practical terms instead of the abstract compartments:

1- Creation permitting understanding (impossible by definition)
2- Sub-Creation permitting understanding (everything we

can know and make)
- Can also be called Repurpose or Recycle permitting understanding

How can this be relevant to the artificial consciousness discussion? Very much so. If consciousness trans-compartmental or Creation permitting understanding is impossible and out of the question, then what options are still there to get to artificial consciousness? This leaves artificial consciousness with no option other than to look for a short cut and get it done without full understanding (Compartmental Sub-Creation or Repurpose permitting understanding).

How might that look like? As explained above, this no-short-cut sequence cannot happen: <u>reality elemental substrate then matter and energy then atoms then molecules then DNA then consciousness (this needs the impossible "creation permitting understanding")</u>. There is an alternative as we might be lucky here. DNA and RNA might be and seems the most ideal candidate out of the above sequence stops to be the short cut we are hoping for. Using the DNA shortcut allows us to drop the following components: <u>reality elemental substrate then matter and energy then atoms then molecules</u>. We can cut in line and starts the pursuit of consciousness with this shortened sequence: <u>DNA then consciousness</u>.

What are the possible strategies to benefit from the DNA and RNA short cut? This may take one of two forms:

1- Exact duplicate strategy (main road path):
It is conceivable to one day be able to take what is available to us and repurpose it into DNA and RNA sufficiently well that it is functional and get to consciousness. This does not appear to be in keeping with the spirit of what most people understand "artificial intelligence" or "AI" generated consciousness to be about. The sequence here looks like <u>DNA then consciousness</u>. The word artifi-

cial here is closer to the meaning of artificial we use for the practice of "artificial insemination".

2- Mimicking strategy (detour path):

This is where the idea behind DNA and RNA codes is used as non-strict guides but rather role models without the construction of DNA or RNA. This might be the route that AI may need to take. The sequence here looks like <u>DNA inspired code then consciousness</u>. This might take the form of DNA digitization then implanting the DNA concepts in machine code. We can call this the "anti-cyborg" where the machine is biology augmented. The concern here is that this has never happened before and seems forced but still these concerns do not represent a concrete absolute argument against AI consciousness.

Three things need to be satisfied (State B in the below illustration) for us to be able to reinvent life and consciousness through this strategy:

a- The consciousness enabling part of DNA and RNA are purely linguistic or code. (not State C in below illustration)

b- If 1 is true, that code carries all the essential information for life and consciousness.

c- That code or language is governed sufficiently by rules and conventions that are within the limits of the aspiring producer (us or machine). (not State A in below illustration)

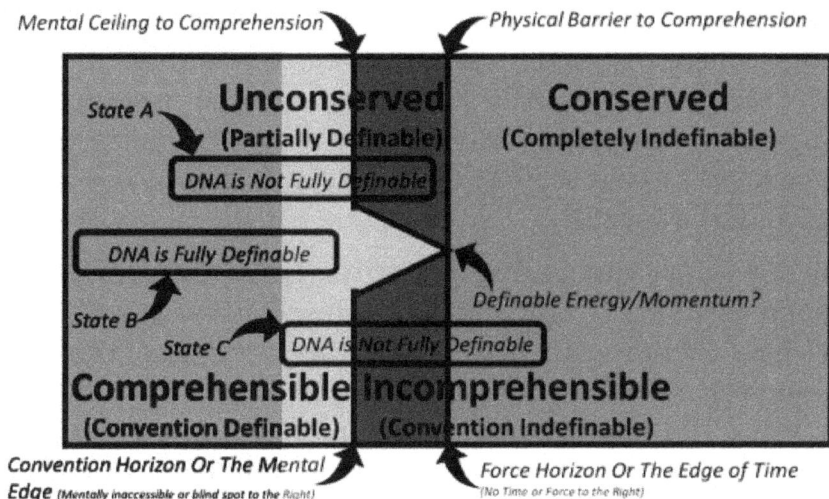

Mental Ceiling to Comprehension — *Physical Barrier to Comprehension*

Unconserved (Partially Definable) | **Conserved** (Completely Indefinable)

State A

DNA is Not Fully Definable

DNA is Fully Definable

State B

State C — DNA is Not Fully Definable

Definable Energy/Momentum?

Comprehensible (Convention Definable) | **Incomprehensible** (Convention Indefinable)

Convention Horizon Or The Mental Edge (Mentally inaccessible or blind spot to the Right) — *Force Horizon Or The Edge of Time* (No Time or Force to the Right)

Is there a potential catch to strategy "2"? Things are looking up for AI consciousness as we may have found a loophole to the humanly impossible trans-compartmental and creation permitting comprehension problems. It appears at first glance that we may be in good shape except for one catch and that is maybe the mighty brain happens to be a Mr. Magoo grade obstacle instead of a dependable enabler. Which is not entirely out of the question. DNA and RNA are non-mental precursors to biological consciousness. Therefore, DNA and RNA are blind to and maybe unimpeded by the mental ceiling to comprehension (the catch). This gives DNA and RNA a potential hypothetical advantage over mental languages (e.g. English, Chinese, French, German, Spanish, Arabic, algorithms, sets, numbers, mathematics, algebra, binary code, programming languages) in becoming a precursor to consciousness.

44. Winning the lottery to AI consciousness

What does the "betweenness model" or "conserved-force model" of reality say about the above short cut approach? The above sub-categories of comprehension (Compartmental, trans-compartmental, creation permitting, sub-creation permitting, ...) are inspired by the "betweenness model" or "conserved-force model" of reality and the proposed "betweenness analysis" and "conserved analysis". But the model permits one more possibility that seems to be easily overlooked and can be lost in all of the details.

Let us not forget that reality generated consciousness before. Although we failed to find another example of reality doing so somewhere and somewhen other than on earth, there is nothing fundamental we know that can show that it could not do that again. What if a serendipitously naturally produced new consciousness happens to take root in a machine. Would that still be called artificial intelligence thou AI had nothing to do with the appearance of the consciousness? AI being part of reality is a participant in a grand lottery game like everything else and might happen to win the consciousness prize by coincidence.

45. When is 20/20 comprehension possible?

The preceding segment talked about blind spots, so it is felt appropriate to say something about where comprehension might gain a deep or full understanding.

Conserved: No

Non-Mental Forces: No

Mental Forces (without definable convention):   No

Mental Forces (with definable convention):   Maybe

Trans-Compartmental: No

Compartmental: Maybe

Creation Permitting: No

Sub-Creation Permitting: Maybe

46. Impossible to simulate

(Disclaimer: Without knowledge of the conventions of the non-human participants in this segment, the discussion is deficient, non-scientific, and untestable.)

In the "betweenness model" or "conserved-force model" of reality it is implausible for consciousness to be a simulation.

A few words about what might be meant with simulation. It is a computer program, alien, or demon generated experience or simulation from outside the brain of an individual who is not able to tell if that experience is coming from the physical world or the experiences are manufactured sensory and emotional inputs indistinguishable. (there is also a variation in that the simulation instead of creating the immersive deceptively difficult to notice experience, the simulation produces the human consciousness completely within the simulation)

Nick Bostrom argued that computer simulations of humans will be done often and refined with each simulation. If that continued refinement with trial and error is coupled with increasing computer power a point will arrive when it will have sufficient fidelity to simulate consciousness, then he went on to say:

"It is then possible to argue that, if this were the case, we would be rational to think that we are likely among the simulated minds rather than among the original biological ones"

Why are simulations even thought of? These simulations are conceptually permittable because there is no clear pairing in the currently available understandings of reality between consciousness and reality or matter. There is a gap in the experience of the relationship. Of related to the mind-body problem.

Is there an understanding that is more immune to these thoughts? These thoughts are not as big of a problem for the "betweenness model" or "conserved-force model" of reality. Here, experience and consciousness are interactive forces bridging in-between fundamental conserved elements. In this model, consciousness is a force, a conserved element associated force. Here, the pairing is fundamental and is relatively immune to be simulated. Without the conserved elements, there is no force. A computer simulation stems from a program or code (not conserved elements). In the "betweenness model" a program or code is a force. A force (code or program) cannot be used to make another force (consciousness). A simulation must repurpose non-force conserved elements to get to a consciousness force. If this happens it is more fitting to define it as recycling or repurposing of consciousness and seems to violate the spirit of what most of us imagine a simulation would be.

This reality set-up (betweenness model) seems more immune to prank external simulation by a demon or alien, but the model seems to build an even more formidable wall that should not permit the fabrication of consciousness by a computer.

47. The simplest and precursor of self-awareness and consciousness

Scenario 4 (See last illustration in segment number: 26), where a conserved elementary entity applies force to itself is the simplest form and possibly the precursor of human self-awareness and consciousness. Scenario 4 is a state of a self-applied force that stands at one end of the spectrum of self-awareness.

A brain's consciousness on the other hand stands further away on the spectrum arguably at the other end of the spectrum, where an enclosed much more complex system applies force to itself.

Consciousness and self-awareness can be defined as the self-application of force.

48. For the sake of the AI discussion, let us do some quick math

(This segment is included primarily for entertainment purposes. There is a clear departure from being methodical. Simply put it is a silly few paragraphs, so don't take it too seriously)

Is computation the proper unit to compare brains with computers? It is a reasonable one for computers, but it may not be so for the brain. Maybe, we can introduce a better measuring stick and call it the betweenness moving parts of consciousness metric.

Now, for fun we will compare apples with oranges, so don't take this too seriously. We will compare computer computations with a made-up consciousness betweenness moving part metric.

Consciousness betweenness moving parts: The "betweenness model" or "conserved-force theory" suggests a force between every conserved element. Let us look at these moving parts a little closer:

1. Force degree: This force is varied in quality and degree (no exact figure is available but probably astronomically high). But for the sake of discussion let us consider it a unit of 1.
2. Force time scale: The stream of consciousness is near-continuous and never the same (a time unit scale is not exactly known but probably small). But let us assume that the mind activities only allow a force/relationship connection once every 150 milliseconds.
3. Nodes: The number of conserved elements in a brain will never be known (no exact figure is available

but probably astronomically numbers of elements are needed to make each atom). But for the sake of discussion let us use the number of atoms in a brain 1.4 x 10^26 atoms.

4. Not all the atoms in the brain are neural tissue. If we take a 3:1 ratio of non-neural to neural tissue. That leaves us with neural tissue atoms count of 3.5 10^25 atoms.

Then we can apply this internet posted formula for the number of connections between nodes or points:

$$x = n (n - 1) / 2$$

From this silly math we can see that:

1. The floor estimate of the consciousness moving parts comes to be around (4.1×10^{51})
2. The ceiling estimate of the consciousness moving parts is (more)[#]

Let us assume that this is the brain equivalent of a FLoating Point Operations per Second (FLPOS). Now we can compare apples to oranges:

- The summit supercomputer does 144 x 10^{15} FLPOS (less than the brain)

- Some suspect 1 x 10^{23} FLPOS may be feasible in 2030 (less than the brain)

Some estimates that are refuted with our above highly accurate calculations:

1. 257.6×10^{24} is the estimated computational power required to *simulate* 7 billion human brains in real-time.
   a. This might be a gross under-estimate. (One brain stands at 4.1×10^{51})

2. 4.4×10^{27} is the estimated computational power required to simulate all humans that have ever lived: approximately $(1.2 \pm 0.3) \times 10^{11}$ human brains in real-time.

    a. This might be a gross under-estimate. (One brain stands at 4.1×10^{51})

3. 4×10^{48} is the estimated computational power of a Matrioshka brain, where the power source is the Sun

    a. A Star level power supply is not enough to generate consciousness if we believe the " consciousness betweenness moving parts" metric.

[#]: 4.1×10^{51} x (astronomic number of conserved elements per atom) x (astronomic number of brain calculation time units per second) x (astronomic number of interaction force variations) + Scenario 4 type force might add more complexity (See last illustration in segment number: 26).

49. Boundedness vs Un-Boundedness

Everything cannot be everywhere. Things of reality do not have only one boundary. Some things occupy a limited range and other things that seem to have an extension where an end cannot be observed.

Examples of range estimates from small to large forces according to "distance" are:

1. Weak Nuclear force at a range of
    a. 10^{-18} meters
2. Strong Nuclear force at a range of
    a. 10^{-15} meters
3. Human consciousness is limited to the skull or about
    a. 0.15 meters

4. Space with an unknown limit. The observable universe is estimated to have a:
    a. Diameter of 93.016 billion light-years
    b. Mass of 1.5×10^{53} Kilogram
    c. Volume of 4×10^{80} m^3
5. Electromagnetic with unknown limit, assumed to be limitless (incidentally impossible according to the "betweenness model").
    a. Infinity
6. Gravity with unknown limit assumed to be limitless (incidentally impossible according to the "betweenness model").
    a. Infinity

Examples of range estimates from small to large forces according to "time" are:

1. Weak Nuclear force: W and Z bosons half-life of about
    a. 3×10^{-25} s
2. The human consciousness force is limited to a lifetime of about 70 years. But consciousness is a streaming composite force with an unknown time scale, but let us assume each mindset lasts
    a. 150 milliseconds
3. Electromagnetic force has an unknown limit. Photons are stable. Let us assume incredibly old.
    a. Infinity
4. Space force has an unknown limit. Let us say the same age of the observable universe
    a. Infinity
5. Time force has an unknown limit. The universe is about 14 billion years old and counting
    a. Infinity
6. Gravity force has an unknown limit. Gravitons are stable. Let us assume incredibly old

    a. Infinity
7. Strong Nuclear force is very stable (Gluons are stable) and lasts billions of years. It may be older than the observable universe
    a. Infinity

For the sake of argument let us remove the weak nuclear force from the list, for no other reason other than it doesn't go with the narrative and it may be an outlier. Sorry, this segment is not about the accuracy but rather about general concepts and ideas that were part of a stream of thoughts during some un-carful free association like-state. Also, precision in the discussed topics are beyond the capability of the writer and probably go beyond the capacity of several lifespans of better men and women.

So, after the weak force omission, we are left with the following observation: the human consciousness force is on top of both lists. The other forces are vastly larger and less bounded. Therefore, we will use human consciousness as our poster child of the most range confined force in reality, for later calculations. And there is also, a poetic fit with what Protagoras of Abdera said:

"Man is the measure of all things"

Next let us ask ourselves: Where did that quality of having boundaries come from? Is it because there is a factor that forces the intrinsically unlimited thing to spread out or is there an intrinsic factor that limits the bounded thing from spreading? The best (but obviously cannot be the only) explanation that comes to mind is simplicity vs complexity. Purity and simplicity aid spread. Complexity limits spread. In other words, Purity and simplicity favor non-locality. Complexity enforces a locality. Let us try to make the concept easier to grasp in two ways:

1. Abstract argument: Let us take another pair and correlate it with the complex-simple pair. That other

pair is difficult-easy. Complexity more closely relates to difficulty and simplicity more closely relate to ease. It is expected of difficult things to be less abundant and therefore less widespread. The opposite is expected for easy things to be more abundant and more widespread. It follows that this carries over to complex (difficult) and simple (easy) things.

2. Example argument: Car factories are more complex than cars. Cars are distributed more widely. Medicine is more complex than raw material. The raw material is spread out more widely. A post-graduate degree is more complex than a graduate degree. A graduate degree can be found more abundantly. Stars are more complex than matter. Matter can be found in stars and beyond them.

Can this be predictable? To be determined, so maybe to some extent. It seems that the complex-simple quality does not perfectly predict boundaries, but it is sensible and true to some limited extent. We will assume its predictability is reliable for the sake of our methodical, air-tight, and profoundly serious discussion.

50. Locality coefficient

For the sake of discussion let us make some more unsubstantiated assumptions. What if we assume that boundedness or locality is predictable? Let us also assume complexity is due to force type or flavor variety? While we are at it, let us assume that the relationship is inverse and linear between locality and complexity. We can put that relationship in a formula as follows:

Complexity (force types) = 1/Locality Coefficient

&

Locality Coefficient = Index range/Reference range

51. Fundamental forces galore

Since our force locality posterchild champion is the human consciousness, let us calculate the totality of realities force variety (diversity), using two framework ranges (distance then time):

According to distance:

Index Range (Brain Diameter)   = 0.15 meters

Reference Range (Observable Universe Diameter)

= 93.016 billion light years

= 8.79999305638e+26 meters

Reality has a force variety   = 1 / (0.15 / 8.79999305638e+26)

= 1 / 1.70454679951423273216098678268995 e-28

= 5,866,662,037,586,666,666,666,667 Force flavors

According to time:

Index Range (assumed Time for each Mind Set)

= 150 milliseconds

Reference Range (Observable Universe Time)

= 14 billion years

= 4.41504e+20 milliseconds

Reality has a force variety   = 1 / (150 / 4.41504e+20)

= 1 / 3.39747771254620569689062839747777 e-19

= 2,943,360,000,000,000,000 Force flavors

Of course, these calculations are flawed but the purpose here is to show that there might be more diversity in the type of forces of reality and according to the "Betweenness model" it is unlikely that the fundamental forces are limited to 4 only.

For fun, let us assume that this methodology is valid (which it certainly is! Wink wink!) and let us see what the brain size should be according to the diversity of forces that the contemporary understanding of reality is (current physics).

Reality's diversity of forces in contemporary understanding = 4 force flavors

Then the locality coefficient of the brain = 1/4 = 0.25

Projected brain diameter = 0.25 x 8.79999305638e +26 meters

= Too Big for The Skull

In other words, for the contemporary understanding to be right our heads need to be bigger than the milky way.

52. Reality In-Touch Index

Sorry, but this and the next segments will not be flattering. We are not doing as well as we think we are.

We can propose a reality in-touch index with the following formula:

Reality In-Touch Index = Established Fundamental Forces / Projected Force Variety

***

Reality In-Touch Index = 4 / 5,866,662,037,586,666,666,666,666,667

= 6.81818719805693092864394471307595e-28

= <1% in-touch with reality

= Not close to truth

= Opportunity for discoveries>>>Discoveries already made

Report Card Grade   = F with ample room for improvement

53. Reality Out-Of-Touch Index

We can also propose a reality out-of-touch index with the following formula:

Reality Out-Of-Touch Index = 1 / Reality In-Touch Index

***

Reality Out-Of-Touch Index = >99% out-of-touch with reality

It is hard to take these numbers seriously, and you should not (using the word "hard" is being generous but the more appropriate word is "impossible"). They are not meant to be sensible in the traditional way we think.

Nevertheless, let me try to take a jab at explaining what is going on. There is one concept that we often choose to overlook, that may make this sensible. No two things are alike. There is more variety in reality than we can imagine. Our poor report card grade reflects that we are not accounting for the omnipresent "uniqueness" of individual conserved elements or anything that reality is made of, and with that comes the "uniqueness" of forces that relates them together.

Our human existence as it stands is based on layers upon layers upon layers of modeling, description, codifying, and classification all erected on a presupposition that "UNIQUENESS" is limited, while uniqueness is a pervert that does not leave everywhere, everywhen, and everything untouched. Uniqueness cannot get its handoff of anything. Hopefully, the above disappointing numbers should be more believable now. It is supposed to be that way if we account for uniqueness fully.

54. Conserved element

variety estimates

We can go backward and use the force diversity estimates to estimate the conserved element diversity and variety.

Let us revisit our node equation, (and assume it is applicable, it does not account for possible node un-similarity, but we will ignore that):

$$x = n(n-1)/2$$

Solving that equation with an x = 5,866,662,037,586,666,666,666,666,667 results in an (n) representing a highly speculative rough apparent estimate of conserved element flavors or types equaling

= 76594138402274 conserved types or flavors.

For perspective let us compare it to other flavors of physical reality:

    a. Conserved flavors: 76,594,138,402,274
    b. States of matter: 4 (maybe 20 if you count nonclassical states)
    c. Chemical elements: 118
    d. Elementary particles: 61

55. The totality of what reality is working with

Reality in total works with a diverse variety of elemental ingredients that can be defined by an inaccurate estimate of conserved element flavors of maybe 76594138402274 and force flavors of maybe 5,866,662,037,586,666,666,666,666,667.

56. Interference effect

The "betweenness model" predicts so many forces that are hard or impossible to detect. If people believe the hypothesis carries weight, then some smart people will be looking for these still unclaimed forces.

Here, screening for interference effects between the forces might be revealing. There may be phenomena that have unexpected augmentation or diminished amplitude due to an unnoticed interference-effect. Interference mining laboratories might be a thing to map out hidden forces. We mentioned elsewhere of some suspected examples, such as stock prices, jokes, and pricing items 9.99 instead of 10.

Admittedly an attempt to find some was not successful. This attempt rested on cross-reference stock, gold, oil, other prices with different currency denominations to try to uncover an interference pattern. The variable and moving part were exceedingly difficult to control for without the proper set up that requires reliable data of prices matched with reliable exchange rates and the computing know-how and power to perform the data mining exercise.

57. How may the base SI units look like if this book carries any weight

Two but possibly more units will be dropped from the list. There may be no further need for a meter or second as base units. Space and time will be forces instead.

58. How do people decide between a dimension or a force? (Left shifters vs Right shifter)

Gravity to some investigators is a force (classic physics). Then other times other investigators would say it is not (relativity).

Then there are other times where other investigators say none of the fundamental forces are forces (string theorist).

It is fair to state that force vs non-force designation is whimsical.

They all belong to the same family of things, their divisions are artificial. Exactly like the borders between countries are make-believe. "Betweenness" is the only thing that can unify them. They all share the quality of "betweenness" overtly or covertly.

What does not matter in this discussion? The name does not matter. You can call them whatever you want. If force makes you feel good, go with it. If dimension makes you feel better, go with it. The betweenness model gathers all of them under the umbrella called "forces".

What does matter in this discussion? Stop artificially dividing them as if there is a deep meaningful divider. Make up your mind. Make them all forces, dimensions, reference entities, or other.......

Let us go back to the table we started within the beginning (see below). This tug of war game of going back and forth about what is a force and not is an admission of our human inability to decide where to draw the "right" black column (see illustration). Some want the black column to stay put where it is. Others want it to move a little to the left and kick out gravity as a force by doing so. Others want it to move the divider all the way to the left and kick out all forces by doing so. The betweenness model of reality argues to move it completely to the right.

World View	Preferred Position of The Right Black Column
1. Classic Physics:	Stay put
2. Relativity:	Move a little to the left (left shift strategy)
3. String theory:	Move completely to the left (left shift strategy)

4. Betweenness Model: Move completely to
   the right (right shift strategy)

[The betweenness purist will not call this right shifting instead the purist will see the right column as a stealthy example of a force that just should join its sisters on the right and left of it, and be a black dividing column no more]

Group	Group 1		Group 2	
Sub-Group	Associated with a Conservation Property	Forces	Others (Non-Mental)	Others (Mental)
Examples	Matter/Energy/ Momentum/ Charge/Angular momentum	Gravity/ Strong Nuclear/Weak Nuclear/Electro-magnetism	Spacetime/ Life/ Genes	Abstracts/ Ideas/ Language/ Consciousness/ Knowledge
Constant in Total	Yes	No	No	No
Standalone	Yes	No	No	No
Changing in Total	No	Yes	Yes	Yes
Creatable	No	Yes	Yes	Yes
Destroyable	No	Yes	Yes	Yes
In-between	No	Yes	Yes	Yes
Interactive Entity	No	Yes	Yes	Yes
Direction	Mostly Scalar	Yes	Mostly Yes	Yes
How may it cause change	Becoming	Being	Being	Being
Symmetry analysis	Intimate Association	Mixed / Mixed	Mostly Not symmetric	Mostly Not symmetric
Pairing with reality	Unsure	Unsure	Unclear	Unclear
Mind-Body Problem	Unclear	Unclear	Unclear	Unclear
Simulation	Unclear	Unclear	Unclear	Unclear
Magnitude	Yes	Yes	Somewhat	Yes
Equilibrium	Not sure	Applicable	Not sure	Not sure
Dimensional analysis	Confounded?	Confounded?	Confounded?	Confounded?

59. We might have a frame of reference problem. Can we talk about it?

Has anyone asked themselves what is the frame of reference for the frames of reference?

In our current human knowledge, the frame of reference is an important issue in any system of understanding. These frames of reference are hotly debated. Some have earned the designation of a preferred frame of reference and others have not been as lucky. In some system they are fixed (as in the Galilean frame of reference), other systems used to refer to the aether (as in the classical or Newtonian frame of reference), some systems had trouble fitting in the accepted frame of reference (e.g. Electromagnetism system compatibility struggles with the Newtonian system), still, others used the observer to relate everything to (as in relativity), then we get into concepts such as an inertial and non-inertial frame of references which adds a layer of more complexity.

What do some of these have in common? Some seem as reactionary to the dominant thought atmosphere rather than intrinsically grounded (they are not reality). They seem to spring from a preceding, hopeful, or aspiring understanding. A theorem gives birth to the frame of reference. They are defined and are asked to function as an impartial witness or judge in a court preceding when the plaintiff created the whole court, furnished it, and hand-picked that witness or judge. If this sounds familiar, it is. It is our friend the Texas sharpshooter fallacy again. The frames of reference do not seem intrinsically grounded (they are not reality).

What else do all of these have in common? They are all unconserved forces. They are effigy products according to the "betweenness model" of reality. It is less ideal to use a varying entity as a frame of reference. Perhaps a more ideal frame of reference is a non-varying frame of reference or namely "conserved entity". Which does not suffer from the variation problem and does not suffer from the Texas sharpshooter fallacy problem too. This potential frame of reference seems intrinsically grounded (it is what reality is made of).

We can paraphrase a famous quote (Man is the measure of all things) to:

*"**Conserved Elements** are the measure of all things"*

60. Foster Parents Wanted

Due to no fault of its own, the "betweenness model" of reality was left at my doorstep. Out of the millions of potential better minds, it could have ended up with, it got shafted in the luck department with me. I tried my best to showcase the idea but felt that the idea needed more.

Therefore, the idea is put up for adoption because it deserves a better parent. Please step forward and adopt this orphan idea and raise it as if it were your own.

CHAPTER 4: PRE-"BETWEENNESS MODEL" MIND-SET TOPICS

61. Chapter Explanation:

This chapter contains several random half-baked ideas that were leisurely compiled in a personal thought diary. They are included here for the following purposes:

- They are referenced extensively in chapter 2 as special case or narrow topic arguments of model breakdown.
- The ideas in this chapter describe the mindset before the conception of the betweenness model of reality
- Some of the ideas seem to be novel and might be worth mentioning in and of themselves independent of the books general theme
- Some might have an entertaining or upsetting value.

62. A law of understanding:

Life's secrets are to minds as laser points are to curious cats. Similar to an endless game of hide-&-seek, where the hidden remains so for the most part. There is an endless quantity of ignorance that is created with every knowledge we discover, it seems. What if unknowables are not conserved and happen to behave like space in-between knowables, not unlike an expanding universe where matter is being thinned out with rabid creation of un-conserved space that comes from who knows where and fills more of who knows what? Feel free to call this the "Ignorance Accelerated Inflation Model". Absurdity at its finest.

Could understanding just be waiting for its "Hubble"? In all directions we look at the heavens of what we know, we see an astronomic number of shining points of knowledge in a sea of dark ignorance. Over the millennia, why haven't the heavens turned completely bright? Intuition tells us that we have progressed a long way in making the skies less dark. It seems safe to accept that knowledge has grown.

Is it, however, safe to assume that ignorance is in a zero-sum relationship with knowledge, where the expansion of one is coupled with the others shrinkage? It is not unreasonable to propose that with each shiny star of knowledge we discover; multiple questions are simultaneously created crowding the space between points of knowledge and causing what an equivalent of red shifting for the stars might be. A shadow of doubt about realness is cast on anything that is not conserved as we understand it. The realness of knowledge and ignorance is therefore debatable as we understand it.

Later and Post-Model Afterthoughts:

1- The realm of the "conserved" (in the betweenness model) in a way has similarities to the realm of the "forms" of Plato. Everything real is of conserved things and their relationships. Everything other than the conserved must be molded by, anchored to, and grounded in a conserved entity or it is nothing more than fantasy. However, it needs to be stated that there is a fundamental difference, where this writing demonstrates a clear connection and relationship between the world we experience and the conserved, which no one has managed to do for the Platonic forms to the best of the author's knowledge.

2- It is no secret that this writing questions the realness of spacetime as it is understood today, but it would be disingenuous if that is stated without some reservations and reluctance. It seems that spacetime as understood today cannot be completely dismissed as it might find validation by being anchored to, molded by, or underpinned to the conserved through "Noether's theorem", where time is suggested to be related to energy and space is suggested to be related to momentum.

63. Being there:

A process membership often systematically neglects the non-member. For instance, mentation often systematically neglects the non-mental. Take non-mental knowledge not meeting the definition of knowledge as an example. Perception often systematically neglects the imperceptible. To react is to notice. Un-noticeables do not illicit reactions. Passing judgment is a reaction. All judgments are blind to un-noticeables and therefore un-noticeables are un-represented. Representation could always be wider because notoriety has limits. Therefore, all judgments are prepared from a prejudiced sample.

64. Reported purposes are cases of mistaken identity of existence for purpose:

Seeing purposefulness where it is not, has consequences that are far-reaching and shape our existence profoundly. Where does such perceived purposefulness come from?

Are your surroundings all that there is? Our intentional interactions are limited to what is noticeable. Noticeable stuff must not cease to exist for us to notice them. Things that cease to exist are not noticeable. Not ceasing to exist is common to all inputs. It follows that all inputs must have a past, therefore possessing a quality of history, therefore, having the appearance of a direction, therefore, giving the impression of purpose.

We can see that it may not be a stretch to imagine that some of the countless things with the quality of "not cease to exist" might be mis-assigned to its closely related cousin quality of "not without intent". We can also see that it may not be a stretch to imagine that some of the countless things with the quality of "cease to exist" are unlikely to be impersonated by the quality of "not without intent" because they are not assigned a history quality and are not seen to possess the appearance of directionality.

If we were made aware of the unnoticeable, it would likely flood us with inputs that would render all other inputs hardly noticeable, leaving us to perceive a tidal wave of "the exist ambiguous". It would be ridiculously hard to imagine a mind that can conclude purposefulness out of anything in that situation.

Therefore, things that we interact with have apparent purpose because things that cease to exist are not there for us to notice, therefore, seeing purposefulness in things may occur with spontaneity often unjustified.

Similarly, perceiving reality obligations occurs with spontaneity and is a one-sided presumptuous hope-based assignment. Reality's lawfulness is premised on the presumptuous hope it is inconsistency free.

When all you have are mind caped laws, everything looks lawful within the said cap. The mind may not have the capacity to account for the lawless, in pretty much a similar way the eye cannot account for the invisible.

Later and Post-Model Afterthoughts:

1-   What do memes, institutions, living, and non-living beings all have in common? A storyline. These things have a starting point and a point further along, which might or might not be an endpoint. In between, there is a story with a direction.

2-   If not purposeful or intentional then what? The odds game might have as much to do with this than purposefulness. It is how natural entities last (produce in big numbers for the few to survive). Most of what we notice has beaten near infinitely small odds for you to notice them, couple that with us being wired to be "purpose first assumers" rather than just "chance first explainers" and you end up with seeing purpose in things where there might be none. Design on the other hand diminishes the luck factor. If great odds were not overcome, purposefulness is more reasonable to consider.

3-   Process inefficiencies are common and commonly go unnoticed, while the small materialized result commands most of the attention. For example, 99.99% or something close to that figure of all species has gone extinct but most of our attention is fixated on what is available. Also, despite the vastness of the universe, we are only aware of one place with life which also commands much of our focus. Ideally, our valuation system and expectations ought to factor in the reality of inefficiency. We ought to be more balanced and account more for things that did not materialize and perhaps less for those that did.

4-   Seeing intentionality and purpose in things when there is none can manifest in several ways. Take this quote for example that points to seeing purpose in somethings devoid of purpose: Hanlon's razor, which states:

"Never attribute to malice that which is adequately explained by stupidity"

It can be paraphrased to say:

"Never say he did it on PURPOSE when it is adequately explained by saying he is just STUPID"

5-   There is a related later segment that talks about the unconserved's propensity to vanish. Skewing our experiential sample size, which probably factors in our tendency to see purpose more often than we ought to.

6-   We can also be victimized by hallucinating "motive" and "intention" in a similar way to what befalls us when we see illusory "purpose".

7-   Morning goodness is a hard pill to swallow.

8-   The future is not obligated to repeat the past.

65. The path to truth is never straight and laden with barriers, to name a few:

Truth seekers will never have sufficient time, number-crunching power, nor enough words. Perhaps the greatest of all barriers is the lack of a direct path to truth, where ideas travel a sea of prosecution before deemed innocent. It goes without saying that not all truths make it across.

Later and Post-Model Afterthoughts: Surrendering brain autonomy might overcome some of these barriers, where brains join forces to simultaneously work on problem-solving instead of functioning autonomously and independently of each other.

66. Keeping a sincere face is easier when you don't know you don't know:

Certainty is not possible but often claimed, and the same can be said of truth.

67. No quitting, embrace being an underdog:

Time swelling, spooky action at a distance, quantum tunneling, and material appearing and disappearing into nothing and all future cousins or the like, often illicit people to say that is weird. Reality is only weird because it's not understood. It's not understood because we lack a capacity or two. Intuition, reason, statistics, and mathematics have fallen short in taking the weirdness away, but this will not stop us from trying, and try we will.

68. Keep believing:

Reading philosophy shakes the certainty in the knowable. Becoming a philosopher entails designating certainty as a unicorn for all intents, constructions, and purposes.

69. The stage is set for the next great mind:

Physics seems to be out of control with claims such as things being at two or more places at the same time and the vagueness of things being a particle or wave. These notions are probably artifacts due to limitations of applied models of reality including mathematic models. Which are ignorant men's attempt to know reality. A reality that evades being truly knowable.

Later and Post-Model Afterthoughts: The author no longer holds some of these thoughts.

70. Good start:

A wise man once said: When all you have is a hammer, everything looks like a nail. When all you have are particle theories and wave equations, everything looks like a particle or wave, other considerations will have an uphill battle to materialize in our psyche until discovered and accepted.

You will never experience what you don't know, as no prehistoric man has ever feared nuclear annihilation or craved a soda.

To rephrase an H.P. Lovecraft quote:

"The oldest and strongest emotion of mankind is fear, and the least old and least strong kind of fear is fear of the unknown"

71. Quit projecting on to me what I am innocent of:

Reality is under no obligation to be sensible, elegant, or simple. Furthermore, it is under no obligation towards humans to provide them with purposefulness, importance, or/and centrality. Nor does it need to provide comfort, security, and/or hope. People craving what gives them comfort, security, and/or hope is not difficult to understand, and if framed in an entertaining narrative it will boost that idea's wide dissemination. This is where pseudo-truths often have the upper hand.

That and, admittedly, being possibly more practical, perhaps plays a role here to a large extent in giving pseudo-truths an advantage. When a car window breaks, it ideally ought to be replaced, but for some of us, covering it up with duct tape is more practical, as it may be simpler, less time consuming, and cheaper.

This parallels what might be going on in the human psyche when it comes to seeking the truth, it often successfully takes the more practical approach of suppressing the craving to know the truth with a convenient pseudo-truth, as a matter of practicality. Comfort food in a way to ease the pain of not knowing.

72. Untamable and playing by its own order or none:

Reality need not take permission or be confined by limits of language, mathematics, wave equations, quantum mechanics nor the scientific method.

73. Truth martyrs are rare:

People might lead you to believe that truth is sacred that we are lured towards it and it is highly sought after. A closer look leads us to see that, truth is often not a priority. It might be more accurate to describe our loyalty to truth as situational. It often takes a back seat to survival and its pillars, e.g. food, wealth, security, and shelter.

Pseudo-truths that deliver security, true or perceived, often are a more effective lure than true truths in recruiting followers. Comfort in being part of a group pulls us to develop a tendency to gravitate towards group opinion, which can be an effective vehicle that promotes pseudo-truths at the expense of truth.

Additionally, the truth might be handicapped by being outnumbered by pseudo-truths that are vetted by less stingy requirements, while truth requires a higher hurdle to clear before an initial public offering.

Later and Post-Model Afterthought:

1- Truth being synonymous with happiness is a common belief

2- As much as truth is held in high regard, defectors outnumber loyalists to it, is just an undeniable brute fact, which might be the world telling us that truth and only truth is incompatible with life. Claims that truth is not and never was a leading force can also be justified.

74. Wish this one is wrong the most:

Intentionality and purpose might be an illusion as we are more likely to experience what survives and are less likely to recognize what goes extinct.

75. Bold prediction:

Relativity seems to be associated with too much baggage and therefore susceptible to being proven to have unrecognized or overshadowed better alternatives. Specifically, speed/constant c as being always fixed is doomed to be falsified. The thought here is that it is a product of a beautiful arbitrary conceptual problem-solving sleight of hand, contemporary mathematics, and physicist's dogma. A similarly workable world view might be choreographed by a different style of mathematical acrobatics, for instance floating c instead of fixing it and making something else non-relative instead.

Later and Post-Model Afterthoughts:

1-   In contemporary physics: The universe expansion assumes that it is not contained, or a capacious container holds it. In other words, distance and space are created from nothing as a fiat quantity. Remarkably like money printing as a fiat currency. We might have a greater insight into the nature of everything if we can find a description of reality that is not based on distances.

2-   For the fun of it these are unaltered notes to self in my thought diary:

Space might be an illusion
The universe expands into nothing
Space is created from nothing
Space might be nothing
Space might be an illusion

3-   Paraphrasing a famous Elbert Hubbard quote:

"To invite criticism, *expand into nothing,*
come from *nothing, be nothing*"

Original Quote:

"To avoid criticism, do nothing, say nothing,
and be nothing"

76. Bold prediction:

Quantum mechanics notion that things can be in two places at the same time is doomed to be falsified. The thought here is that it is an artifact of limitations in contemporary measurement methods or understanding. Similarly, the Hansberger principle of uncertainty may be a product of measurement limitations.

Later and Post-Model Afterthoughts: The author no longer holds some of these thoughts.

77. What is available is assumed to be what is desired, but what is desired is never available:

A description is a cheap alternative to knowing reality. Reality's unknowability forces us to use its close substitutes of descriptions, pseudo-knowledge, or models.

78. The best language is the one you speak, or is it?

No easy way to say this, but some languages are simply inferior, and it might be yours. If you are like most of us who assume that different tongues capture meaning equally and languages are just different but similarly valid ways to pass and receive information, then look a little closer and you will note that is an impossible assumption. There must be differences in the performance of information carriers.

Later and Post-Model Afterthoughts:

1- English is said to be the language of science. At best this is only partially true and possibly only temporarily. The closest language to be an uncontested language of science is mathematics, and that is limited to the hard sciences and it is not sufficient to not require other languages. There are obvious short failings of English and mathematics. Therefore, this author cannot stop fantasizing about a discovery of a new language that can be used to describe the world better than English and mathematics and by extension reframe the way we think in exciting ways. Could there be a language that does to mathematics what Arabic numerals made possible compared to Roman numerals or what algebra made possible compared to mathematics before symbol use? Could there be a quest for the "language of everything", like there is a quest for the "theory of everything"?

2- If you are looking for evidence and if there are any doubts about the force languages possess, look no further than the lines of stock price support and resistance at whole numbers and multiples of ten. These imaginary floors and ceilings have no relationship not whatsoever with anything fundamentally real about the stock value. Do this thought experiment for fun, without changing anything fundamentally real even a bit, these lines of resistance and support would be quite different if an abstract or language switch such as using Roman instead of Arabic numbers was used. Our Betweenness or Conserved-Force Model of reality (see below illustration) offers an explanation of this phenomenon.

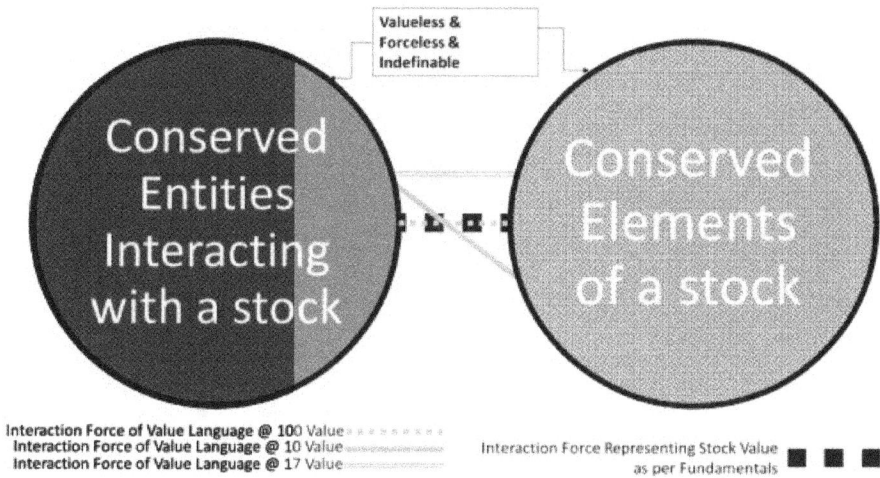

Valueless &
Forceless &
Indefinable

Conserved
Entities
Interacting
with a stock

Conserved
Elements
of a stock

Interaction Force of Value Language @ 100 Value
Interaction Force of Value Language @ 10 Value
Interaction Force of Value Language @ 17 Value

Interaction Force Representing Stock Value
as per Fundamentals

3- The theory of "language tampering" further expands on this topic because the above illustration is a special case example of "language tampering" which is also explained by the Betweenness or Conserved-Force Model of reality.

79. Exposed nature leaves little room for the imagination:

The machinery of science is a wonder buster.

80. Keep telling myself to stop looking for limitations, but they are everywhere:

Painful realization: In the field of AI the Turing test is used to test for human intelligence. It seems as if our intelligence is a matter of other's opinions. How limited we are is embodied in the power of our diagnostics.

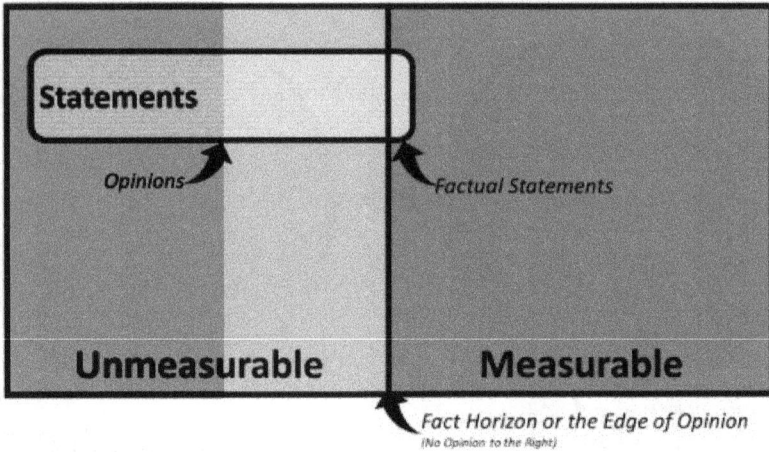

Fact Horizon or the Edge of Opinion
(No Opinion to the Right)

81. Don't judge info by its factuality alone:

Truths may be detrimental, while falsehoods may be helpful. Accident, chance, and coincidence do not provide a full explanation.

82. Uncertainty principle:

Nothing is sayable. All words are a classification of meaning by convention. Meanings are a continuous spectrum, while words are quantized. Capturing all meaning in words is impossible. Meanings are inconstant and words are fixed representation at a single time point of.

Warning!!!! Shameless self-promotion ahead: Alokaili's uncertainty principle states that no meaning can be expressed precisely and indefinitely at the same time. The more precise the expression is the narrower the scope, and the wider the scope the less precise said expression becomes.

83. No terms, No winning:

The only way to completely know reality is to rig the game, by redefining reality into narrower terms, and then it can be grasped. For instance, Berkeley claims that things must be sensed or perceived to be substantiated/exist.

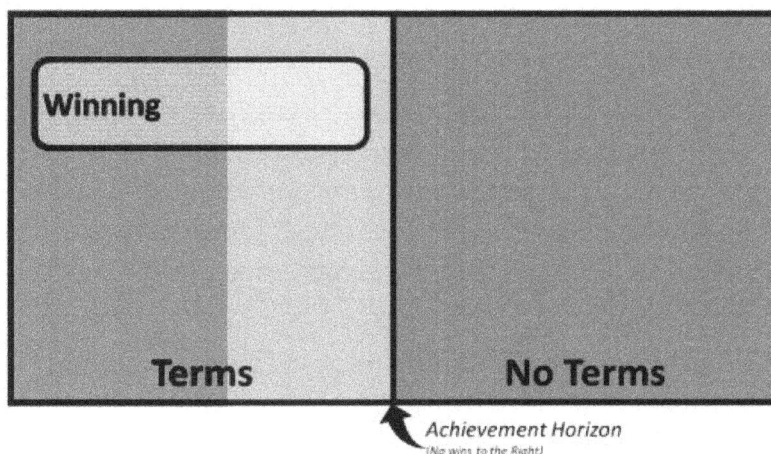

Achievement Horizon
(No wins to the Right)

84. If organisms were numbers, none would be a prime number:

All organisms are composites, and that is not to mean of cells, atoms, molecules, body parts, or organ systems. The collective will of the organism is not limited to what the mind desires, but there are other will forces (genetics, background survival programming, evolutionary influences, ideas) at play, that mix to give birth to a net will force, that drives us.

This collective will gravitate towards a target state that materializes in the absence of physical, environmental, or social obstacles.

A gap can be surmised to exist between this natural target state and the target state which can be hypothesized from deductions of intellectual thinkers, who seem to overrate the sway our intellect and conscious mind commands over the will. This gap is both another example and a partial explanation of the ought/is problem.

Interestingly, you and all of us are a composite of symbiotically cooperative and parasitically elbowing factions, where, out of all the components, the part that thinks and is conscious is the part which cannot agree on what its purpose or goal is. While the other components do not suffer the same dilemma.

These components, that you unknowingly carry with you, can become pernicious without you suspecting it. There are superficially similar precedents, for example, most of humanity did not know of the bacteria they carry in their guts and on their bodies until recently. These passengers are essential for life but can revolt to your detriment.

Word of advice keep the peace (embrace our universal "Innate Stockholm syndrome" we conveniently call self-admiration or self-love) as long as you could, and strategically pick your winnable fights, the "conscious you" cannot be liberated, and the abducting non-you in this dynamic will see you to your finish no matter.

Later and Post-Model Afterthought:

Genes were said to be selfish towards us. The selfishness attitude goes far beyond genes as the same can be said of memes and ideas being selfish towards us too. Where do we, you, and I figure in this dynamic?

You, we and I, are the third wheel in this threesome, the recipient of the neglect and the victim of lack of attention. You, we, and I are the battlefield to these two love birds when they fight. You, we, and I are the matters and beds that host this horny couple when they are not fighting. You, we, and I are their resting place, the arena where they have fun with each other and alone, and yes you, we, and I certainly get stained by the blood, sweat, and other fluids in the whole process.

This exacts a toll in the form of wear and tear on us, and when the time comes where you, we, and I are of no further use they have probably already moved on to other unexpecting host battlefields, mattresses, and beds.

If Richard Dawkins was the sun, it may have written a book titled "the selfish energy or momentum". If Richard Dawkins was a gene, it may have written a book titled "the selfish consciousness". If Richard Dawkins was an AI, it may have written a book titled "the selfish code". If Richard Dawkins was a community, it may have written a book titled "the selfish individual". If Richard Dawkins was an individual, he/she may have written a book titled "the selfish society". If Richard Dawkins was an idea, it may have written a book titled "the selfish gene", oh wait a minute that was written!

85. Wishful thinking:

The history of ideas has plenty of overly optimistic or deceptive notions, such as the "will to power". Things beyond our sphere of influence, overshadow what is within it, rendering us largely not in control. Try willing away your place of birth and parents or even better gravity and time. Wishful thinking is so ancient and abundant, which makes it more common than we can fathom and hard to spot.

86. Kingpins:

Circus animals and we are not unsimilar. Reward triggers vs punishment triggers compel people to follow a scripted life. It is doubtful that circus animals have the option to move freely, they are also trained with reward and/or punishment triggers to execute certain scripted tasks that are monetized. Similarly, something is cashing in on your scripted actions.

A cow's spots do not matter much to meat and dairy lovers, nor is your individualism. Individualism is a sideshow to the main attraction. The actions that most if not all humans do collectively are where the money is. That is where you start your inquiry and you follow the money forensically.

Let's take it even further and not be too egocentric, the better question is what do you do in common with all living beings? Then narrow the circle to what do you do in common with all conscious beings? Then narrow the circle to what do you do in common with all thinking beings? At each level of complexity, there is a kingpin you are working for.

Later and Post-Model Afterthoughts:

1- A confusing personal experience that is difficult to wrap my head around goes like this:

Rarely I am aware of my self during sleep in a confused state where I have no control over my body. I desperately want to take in a breath, but my chest and lungs won't. I picture myself suffocating with my head facing down against the mattress where it is not possible to exchange air. I feel that my limbs are trapped and strained and are not where they are supposed to be and commanding them to free themselves goes nowhere. Terrifying moments of helplessness as this seems to be the end. The initial efforts to will my functions to obey are quickly followed by giving up and surrender to what seems to be the initial steps in the process of dying. Until I wake up suddenly not because I willed myself to wake up but because it just happened. I was at the complete mercy of something that woke me up, something that blocked me from breathing, something that had my head facing down, and something that intercepted all commands to my limbs. Nothing even comes close to this experience of being inside something that was once under your full control but without advanced warning it suddenly no longer is.

The closest experience to this is the moments of paralysis I felt when I fell from a tree and experienced a short-lived spinal cord shock where the hit was such that it switched off my spinal cord and stopped my breathing and ability to move while the fall was not sufficient to switch off my brain and knock me out. This state of uncoupling between me and my control over my body is deeply frightening.
What does it signify? Should this experience be ignored as an illusion? Should it mean body and mind are not the same? And again, who is the real boss or who owns me if my own "voluntary functions" can terrorize me by going on a synchronized coordinated strike?

Before anyone starts pointing out the obvious, yes this phenomenon is well known and the technical term for this phenomenon is "night terrors" but that technical information does not take the odd feeling away, and the questions that were just asked still feel relevant.

2- Individualism might be a delusion.

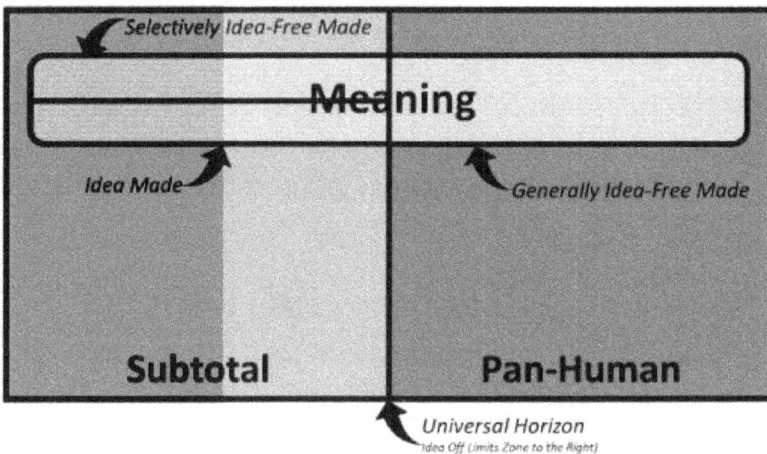

87. Multiple realities don't exist but are undeniable:

It was shocking until it was not, that thinkers and influencers do not have a real handle on reality, nor anyone else for that matter.

88. It is unsettling, anytime a structure of a basic question may need to be revised:

"Where and when am I?" are unsettled questions. Kant's theory of intuition used time and space as examples of fixed metrics, this intuitive claim about reality was shattered by Einstein's relativity which fixed the speed of light "c" instead. Don't believe for a second that this matter is settled, it is likely that fixing c is a mathematical sleight of hand that tortured reality to play along. "Where and when am I?" may not be a proper question after all, and hence the confusion on what may seem to be a simple matter of I am here and now, is not so simple after all and might be improper qualities of a state.

Post-Model afterthought:

The "Betweenness Model" or "Conserved-Force Model" of reality offers a different explanation to the time and space phenomenon, compared to the concepts placed forward by Kant and Einstein.

89. Translation does not mean identical:

All spoken languages have a word for happiness. Are they all equal in representing this emotion? Even if single words of different languages can somehow have an identical representation of reality, having the sentences of different languages be identical representations of reality is hard to believe. Can something similar exist with mathematics since it is a body of languages too?

```
01000011  01100001  01101110  00100000  01110011
01101111  01101101  01100101  01110100  01101000
01101001  01101110  01100111  00100000  01110011
01101001  01101101  01101001  01101100  01100001
01110010  00100000  01100101  01111000  01101001
01110011  01110100  00100000  01110111  01101001
01110100  01101000  00100000  01101101  01100001
01110100  01101000  01100101  01101101  01100001
01110100  01101001  01100011  01110011  00100000
01110011  01101001  01101110  01100011  01100101
00100000  01101001  01110100  00100000  01101001
01110011  00100000  01100001  00100000  01100010
01101111  01100100  01111001  00100000  01101111
01100110  00100000  01101100  01100001  01101110
01100111  01110101  01100001  01100111  01100101
01110011 00100000 01110100 01101111 01101111
```

Post-Model Afterthought:

Look no further than jokes that often lose their humor when delivered in a different language. The "Betweenness Model" or "Conserved-Force Model" of reality offers an explanation of this phenomenon.

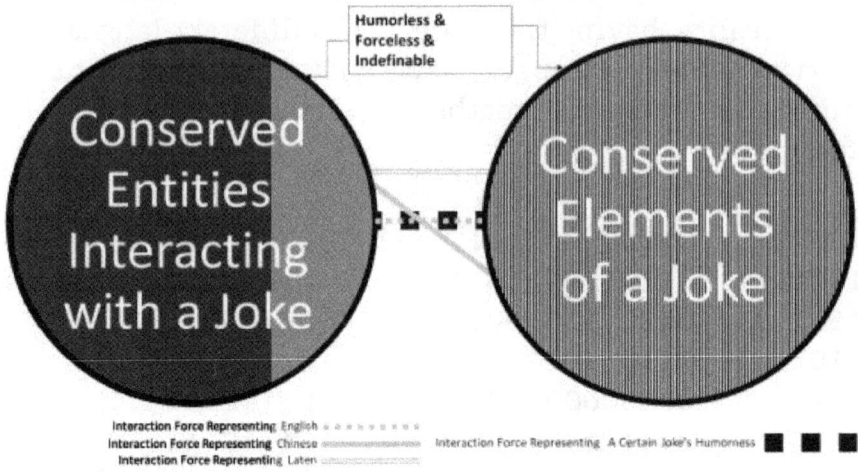

Humorless &
Forceless &
Indefinable

Conserved Entities Interacting with a Joke

Conserved Elements of a Joke

Interaction Force Representing English
Interaction Force Representing Chinese
Interaction Force Representing Laten
Interaction Force Representing A Certain Joke's Humorness

90. Here by myself:

The only non-self reality we can experience has to be independent of time-space and conventions including language. Good luck finding such a thing, I haven't. Which leaves us with, experiences being all models of reality and knowing as a delusional state. The phrase: people are not ready for the truth, is built on a wrong premise, and fall short in a literal sense. Rather people have never experienced truth from the onset, and their readiness is purely hypothetical and untestable.

91. Disagreeable bunch (Part One):

Most people are indifferent to the questions of philosophy, which is precisely where some philosophers suggest people should remain. For instance, Heidegger reportedly claimed that people don't bother with the meaning of being (his definition of what philosophy is) and that philosophy was an obstruction to understanding the world. Ironically, he is a pro-thoughtlessness thinker.

Heidegger is effectively taking a holiday on behalf of philosophy and therefore has possibly staved a philosophical nervous breakdown if we were to believe what a wise man once said:

"One of the symptoms of approaching nervous breakdown is the belief that one's work is terribly important, and that to take a holiday would bring all kinds of disaster"

92. Hegel is wrong:

The evolution of inanimate things gives us the illusion of a will and an apparent consciousness. For example, governments grow, and history has been said to have a will. Hegel might be in the wrong on this matter, by falling in the trap that we are all inclined to walk in, which is counting on what is observable to be all there is.

93. Next gold rush candidate:

What is not perceived carries importance in all aspects of reality. One might predict great insights and breakthroughs if the bigger picture is taken into consideration. Statistically and historically, we have been better at grasping the numerator and under-estimate the denominator, or the other way around, anyhow one is systematically underestimated. Then, how can you give weight to what is perceived if you do not acknowledge the alternatives that did not materialize?

94. Consider it a "Repeat to me what you think I said?" type exercise:

It goes without saying that this misdeed of a book or writing is not the work of a person qualified to philosophize, but this is precisely one possible point of this work. The point is for it to contain an illustration of how an unqualified individual may fair in this regard and to hopefully provide polished minds of philosophy insight into how much of their output registers with our types.

95. Your mind is the ultimate "Photoshop" and abstracts are a popular brush stroke function, possibly unique to its human premium edition:

Abstracts go through a process of conception, then possibly belief, then possibly followed with disbelief. Abstract entities regularly come into existence with spontaneity un-conserved.

They are granted authority by belief and they are stripped of authority and possibly forgotten with disbelief.

This dynamic of building and tearing down of abstracts occur in all of us and at any given moment, our set of abstracts is rarely, probably never, identical.

Abstracts often shape your perceived reality. Imagine a world with no ZERO, and you cannot deny that the same reality feels different. Therefore, there is no single perceived reality but rather a plural of perceived realities.

We can often see mismatches between what humanity collectively perceives as real and how reality is perceived in any individual's mind, partly because they don't have the same sets of abstracts (e.g. Western, Eastern, ...).

It is difficult not to feel queasy because life is simpler when material things are all there is. It is easy to discount the non-material (e.g. abstract) until you realize it is a force to be reckoned with.

Later and Post-Model Afterthought:

1- On initial inspection, it looked fishy that Einstein needed different mathematics other than Euclidean mathematics to justify his theories. Superficially, it resembled a clever exercise of moving the "goal post" (play by different rules, change the conditions of engagement and thereby rigging the game to makes it easier for something to be a winner, successful, or truthful).

Could different languages induce different realities? Could the numerous spoken languages, Roman vs Arabic numerals, decimal numbers, algebra vs classical mathematics without symbols influence our capacity to create abstracts that will spill over to shaping our perceived reality? Does a two-way relationship between us and language exist, in which we manufacture language, which in return colors how we see the world and thereby plays a role in manufacturing our ideas? It is hard not to believe so.

Going back to Einstein, further inspection and listening to experiments about time inflation and light bending makes it more comfortable to not think of what Einstein did as being a hoax (this is not a comment about intention, which undoubtfully are not sinister. It is just dramatizing for entertainment purposes. Also, it is not about singling out Einstein, but these opinions extend far beyond him to the whole practice of language-based learning including but not limited to physics).

In all seriousness, it most likely appears as if Euclidean mathematics broke down and was wrong which made it necessary for Einstein to seek a better tool/language that is less wrong for his purposes, in the form of non-Euclidean mathematics.

2- At this point, it is hard not to play a game of "what ifs?". Such as, what if the uncertainty principle is based on or manufactured by the wrong mathematics?

3- The uncertainty principle is linked to the absence of

a vacuum and virtual particles and energy popping from nowhere. It is also linked to particle-wave duality. It is doubtful that mater is truly wavy or particulate. It is also doubtful if space and time truly exist in the way they are described nowadays but are rather a useful framework to organize our human experiences at the moment until something better comes along. In the "Betweenness Model" or "Conserved-Force Model" of reality, space and time are described as forces.

4- If you are looking for evidence and if there are any doubts about the force the abstract possesses, look no further than the lines of stock price support and resistance at whole numbers and multiples of ten. These imaginary floors and ceilings have no relationship not whatsoever with anything fundamentally real about the stock or business value. Do this thought experiment for fun, without changing anything fundamentally real even a bit, these lines of resistance and support would be quite different from an abstract or language switch such as using Roman instead of Arabic numbers.

5- The theory of "language tampering" further expands on this topic elsewhere.

96. The ultimate milestone:

If you think you know anything, think again, we still are not able to come up with anything solid or generalizable with our senses, augmented senses (or instruments of measurement), reason, and augmented reason (e.g. mathematics and statistics). The highest intellectual milestone would be to break through this seemingly impenetrable barrier to absolute truth, which no human has ever achieved. It is doubtful that the front runners (reason and/or science) will get us there, we may need and await the emergence of a different truth-teller.

Later and Post-Model Afterthoughts:

1- Statements are either wrong or correct, and correct ones never go far enough.

2- Upon further reading, there appear to be special-case examples (for formal systems) of expert support to the "condition boxed dynamic problem" stated in this writing. Gödel's Incompleteness Theorem tells us that Gödel proved that no formal system could tell you the truth, the whole truth, and nothing but the truth. He showed that no formal system can answer some computations.

97. Not all the pens and keyboards could make mental language whole, never to outgrow being an incomplete summary:

Look no further than the language of Hegel to see the limits of language as an emissary of knowledge.

Also, the limitations of language are implied by the existence of many complementary forms. For instance, the existence of non-verbal language implies limitations in verbal language. The existence of emoji implies the same towards alphabet written language.

Another pointer towards the unwholesomeness of mental language is that it is a product of and requires a brain and therefore inherits the brain's limitations.

Further to this point, is that it is curious that knowledge is defined to be mental, ignoring non-mental knowledge that is learned and passed from one generation to another thru the language of DNA and RNA. Regardless of whether you call it knowledge or information, it (DNA and RNA information) undeniably has its language that includes alphabets, words, sentences, and chapters. DNA and RNA go through proofreading constantly, and editing can be both natural and unnatural.

The mighty brain and mental languages comparatively (to DNA and RNA) fall short in several regards such as the:

1. Need to be learned
2. Not transcending cultures and borders
3. Is not comprehensible to all ages
4. Suffers subjectivity
5. Requires consciousness

Later and Post-Model Afterthoughts:

1- It seems that rationalist and empiricist did not account for non-mental knowledge transferred to us through DNA and RNA. Not, sure how that impacts the discussion about the knowledge debate, but knowledge cannot be all from experience and it cannot be all from reasoning and it cannot be all mental.

2- If DNA and RNA information is not an example of a priori information nothing should. Not belonging to the mental club is taken against it of course.

98. Language is a head-scratcher, but I'm glad we have it:

Categorization is fundamentally flawed. Words or units of any language are symbols of categories and are therefore flawed. Practically speaking, every non-abstract thing in life has blurry margins and is never pure but rather falls somewhere along some sort of continuous scale. Therefore, when Hegel makes conclusions such as ideas drive history, rather than economy or instinct he assumes wrongly that ideas are a pure standalone real entity rather than a word that symbolizes and flawdulantly (don't google it. It is not a word until now) categorizes ill-defined impure products of the mind, that for sure are contaminated with different worldly factors (such as economy and instinct). The same applies to other words such as history and spirit that Hegel is also fond of.

99. Not all fantasies have declared themselves:

Many pure claims are a product of a point of view or a narrow perspective. A different observer with a different point of view or a wider field of vision might have overlapping claims with a different level of certainty.

100. We are all haunted by ghosts of cavemen past:

Instinct and feelings are the net output of the mind's calculations, which fill gaps in knowledge, with educated guesses. Its accuracy increases in matters that we share with cavemen and are less accurate when we deal with less primal/primitive matters such as time/space and quantum mechanics.

101. Our understanding is partially satisfactory:

August Comte's developed his stages of understanding: mystical, metaphysical, cause & effect based (i.e. scientific or positive understanding). Mach preached to remove metaphysics from science, which was endorsed by the Vienna circle, which echoes Mill's stand from even before that. This is in contrast to intuition as a source of understanding, including Kantian's a priori concept.

All these stages and divisions can be combined under one umbrella as they belong to one family we can call "Conditional Understanding".

Satisfaction will be withheld until the long-awaited next stage of "Unconditional Understanding" arrives. Will, we ever realize and achieve this satisfactory stage, remains to be seen? It seems impossible unless a grand and radical discovery is made, such as time doesn't exist as described.

Later and Post-Model Afterthoughts:

1- Some understanding gaps that need to be bridged include distorted stimuli before reaching you, our limited perception apparatus, language being symbolic, nothing happening twice, no two things are alike, finite lives, consciousness, and solitary mentation.

2- Truth being synonymous with happiness is a common belief.

3- It is difficult to decide if we should praise or criticize humans on a job well or poorly done when an idea is labeled truthful on the merit of its success or labeled a fallacy on the merit of its failure.

4- Just wondering if a simple message (like we are equal) thou imprecise is what we need. In other words, do we need or want the truth?

5- Some handicaps might be desirable. For example, the capacity to forget or our capacity to feel pain. Do you wish for a perfect memory and a complete absence of pain? Our understanding being handicapped with conditions might be so too.

102. A first, with "fulfillment

pending" status:

It would be the exception to the rule if the mind understood reality. A long-standing tradition will have to be shattered. A single example of a product understanding its creator escapes the author. Attempts by humans so far are not so good, the mind's performance in this regard is honorable but leaves a lot to be desired. Look no further than philosophical truths over the centuries, none are unanimously accepted.

Nothing else seems to care about trying to understand truth apart from us, except perhaps machines, in theory, might one day, in the unlikely scenario that they develop a curiosity.

For machines to be able to accomplish what we haven't, it likely must figure out or stumble on to a novel way to learn, which almost certainly has to overcome many insurmountable barriers for us, such as time, or it will always be living in the past as we do.

103. Asking too much of tools:

Heidegger's view about reflection and engagement with (knowing/observing vs getting the task done, in other words, we are not primarily knowing beings but at a more primordial/basic level we are involved in the process of completing tasks) life has some parallels to the notions expressed in this book about codification/ classification/ categorization/ words and reality, and also the notion that we are using the mind for acquiring knowledge which happens not to be the primary function it was built to do. This is not ideal.

Gently put, it is not unlike using a car's backseat as living quarters and a car's trunk as a closet and personal storage facility.

Bluntly put, using the mind to acquire knowledge is just like asking a car to take you to Andromeda. Arguably, Georg Cantor wrecked his car trying to.

Later and Post-Model Afterthought:

Let us face it, asking a mind to acquire certain truth is like asking Stephen Colbert to be unfunny or wrong. It simply is not going to happen.

104. Disagreeable bunch (Part Two):

"Passion is cancer" from Kant's point of view but is "essential" in Hegel's point of view for the progress of history and ideas.

105. Fact-checking loopholes:

The stealthiest ideas are potentially the most damaging ideas as you are often defenseless towards them. Little else, in this regard, compare to ideas handed down to you by your parents. Not closely related but, since we are on the topic of loopholes, another loophole of fact-checking is when we go to sleep, where fact-checking is suppressed and dreams can get unrealistic, freaky, and weird but still don't feel that way.

106. Being everywhere can be a cloak of invisibility:

All experiences are models of reality. The only reality we can experience must be independent of time-space and convention including language. As mentioned elsewhere, the phrase "people are not ready for the truth" is built on a wrong premise and falls short, but rather people have never experienced truth from the onset.

What did you say? Did I hear you say something! Be careful who you are calling "detached from reality" because so are you and all of us. We only differ in the degree of detachment, but a constant never-ending state of detachment from reality is indisputable. The cloak of its invisibility will come off when it is large enough to cause dysfunction.

107. Why a state of not knowing that we don't know?

Most don't realize we are living a lie or partial truth. This may not be an accident, because who are we to claim or know that a world of mostly knowing what we don't know, knowing what we know, or not knowing what we know is advantageous. For all we know, not knowing that we don't know maybe best for the human condition, as there is no way to know if we and our existence are more suited for the alternatives including the absolute truth.

If we look at all the operations that any living organism might do, most of them are hidden from consciousness. Why isn't it otherwise? We might ask ourselves, why don't living organisms have a freedom of information policy that allows consciousness a look at the mechanics of life. Because of this secrecy, nosy consciousness has erected elaborate spying, espionage, and an intelligence-gathering apparatus where it gathers information to uncover what its own have hidden from it.

108. Wrong us for the job:

Partial obstacles to the final intellectual frontier are the current tools and arrangements: speech, reading, science, and maybe even DNA. Such an objective might only be realized by abandoning humanity all together for a different form or retention of some but not all of what makes us human through a hybrid symbiotic existence with non-human parts.

109. Posterior probing digit vs thermostat:

There lies a difference between a posterior probing digit vs a thermostat, one might be a violation, the other will measure core temperature.

Our history, nearly in its totality, is of people claiming access to truth, when all they uncovered are knock offs. Because of that, it is hard to take people's claims seriously, even to not get offended by them sometimes, or not get outright hurt by them other times. Will we be in a constant juxta-truth state for eternity? Or will we achieve truth realization?

How may that look like if it were ever actualized? It is hard to imagine that we can trust claims of said accomplishment if they occur. Claims are not reliable.

If claims are not good enough, then what can be trusted? It might be reasonable to surmise that signs of such a development have to rely on witnessing its manifestations. It might also be reasonable to conclude that such manifestation will be disruptive. One might suspect signs of approaching the final truth, would include conquering aging (health care and medicine are making a baby step in this regard) relinquishing speech (efficient real-time translation is a baby step in this direction).

In the author's mind, as long as people continue to talk and age, you can rest assured that we haven't found truth. This might be one potential explanation why we haven't been visited by time travelers or have been visited by them but don't know about it because, the need and urge to communicate and be observable is not a post-truth thing.

Posterior probing digits are not a good way to check for a fever, yet that is the state we find ourselves in continuously on a global scale each time anyone f███ f███ us with the "truth" as far as we live in the pre-truth era.

Later and Post-Model Afterthought:

While watching a recorded Eagles vs Cowboys game hours after the match, without knowing the score, it felt live. It was a confusing feeling that raised a question.

What is the difference then between seeing a game live and recorded? Then it occurred that the live broadcast is not live but is also a recording but is not allowed to wait for long before being watched (a few seconds for broadcasts vs hours, days, or longer in the traditional tape or hard drive recorded game).

Let us take this logic one step further. Does seeing the streaming live feel like seeing the game in the stadium? They likely generate similar feelings. Then it occurred that experiencing the game in the stadium is not live but is also a recording but is not allowed to wait for long before being watched (a few milliseconds for stadium watchers vs seconds in live streaming or live broadcasting).

Everything we experience is recorded in some way or another even if we feel it is direct and instantaneous. This resembles our dilemma with truth, we are always juxta-live even if we feel things are live or call/label things live or true.

110. Off label use:

What is the purpose of the mind? Why do we have it instead of not?

If we take a hammer, the purpose of a hammer is to apply blows to things such as a nail which may cause it to be pushed into something else. We know this because, this corresponds to what we see, in which hammers achieve that purpose, and therefore a state of "purpose/achievement concordance" is established.

We can apply this concordance relationship in reverse to guess the purpose of things. For example, a hammer has never once functioned as a source of vitamin-D, therefore, it is not the purpose of hammers to be a source of vitamin-D because there is "purpose/achievement discordance". Similarly, no one thinks a hammer is something we can use to gaze in the stars. On the other hand, we verifiably see nails being pushed into things with hammer blows, therefore, it is fair to state that as a purpose of that hammer, because it does not violate the concordance rule.

We (or at least some of us) are using the mind to know what reality is and find the truth. We have not found either in absolute terms. There is a "purpose/achievement discordance" that allows us to question that the mind's purpose is to find truth and know reality.

It is fair to wonder if not to find truth and know reality then what is the mind's purpose? A good exercise might be to see what the mind achieves the most to guess its most likely purpose.

The brain's apparent drive to seek the truth might not be so different than a mosquito's determination to fly towards the light, a glitch of some sort.

Later and Post-Model Afterthoughts:

1- The phenomenon of insects actively diving into a light source or fire is a fascinating example of a mortal assumption. It occurs because insects assume incorrectly that all light is moonlight (a safe assumption before the invention of artificial light and humans acquiring the ability to light, control, and employ fires). Some insects use moonlight to navigate. Moonlight because it is distant light is practically parallel, but on earth not all light (e.g. man-made artificial light and fires) is distant and therefore it cannot be assumed to be practically parallel if coming from a proximal source. If the insect happens to be near a non-moon light source and it catches its attention, it can make the insect's navigation system miscalculate and feel as if it is flying straight while it is spiraling towards the proximal fire or light source and possibly to its demise.

Our minds use an incalculable number of assumptions, could one or more of these assumptions cause an anomalous miscalculation which sets us on a spiraling course toward the light/fire of knowledge while our minds think it is sailing in the straightest path towards truth, safety, survival, and peace.

2-   A brain is a survival tool that happens to think (for whatever outcome be it positive or negative).

3-   Suicide is so rarely observed in animals and those observed examples don't sound human-like or convincing. So effectively suicide may be practically considered a human phenomenon. It is always fair to accuse ideas of being behind anything animals don't do while some humans do. Therefore, it is fair to accuse ideas of many of the suicides. This suggests that many suicides are a mind-made or an idea-based phenomenon. Could suicide be an undesirable side effect of one or more generated anomalies following the "off-label use of the mind"? Do not expect a plant to commit suicide.

111. Which one is real? Neither:

Matter and energy are the same we are told but are called two different things in spoken language, while they are unified by mathematical language. Both languages are different modeling tools of reality that serve non-identical purposes, therefore, co-exist instead of being substitutes of each other.

Non-identical purposes are why expressions of the same thing could not be the same unless some form of compromise transpired or it happened through a coincidence. For every point of interest, a purposeful expression or more likely exist if permittable by the expressive limitations that sometimes cause somethings to evade an expression assignment and remain unfulfilled because expressive tools are no match to the wider less restricted interest and point of view capacity.

Expressive modeling restrictions have been discussed elsewhere, therefore this segment will be more about interest and purpose.

Once a purpose vanishes a corresponding model is more likely to follow suit. Every purpose will seek a modeling tool or more to satisfy it. No expression or model has ever been created without a purpose. All-purpose is grounded by interest. Interest bread purpose. Purpose drives all expressions. All expressions and models are therefore derivative.

Reality understanding is a human interest, that creates purpose, that seeks a modeling tool, that produces expressions. Reality here is a bystander passively going on its business, free of human qualities such as interest and purpose.

Will any of the modeling tools that are universally confounded by interest and purpose ever be able to truthfully represent an interest and purpose free reality? It seems questionable. But admittedly this line of thinking is not an airtight argumentation that all interest sparked initiatives will be incapable of finding truth, but it makes it one or more steps removed.

However, this line of thinking predicts that as long as we have non-singular human interests, we should not expect those expressions about truth to be any different unless compromise or coincidence intervenes.

If you feel that you've seen similar examples before, so have I. Let us bring up Wave-Particle duality as a suspect concept that uses two different expressions to describe the same thing, because they must have non-identical purposes or they are an artifact of an expressive handicap and therefore, two expressions team up to represent one thing. All multiple expressions of the same thing are a cover-up of our limitations or non-identical purposes. It is highly unlikely that Wave-Particle duality is an exception.

Other pair or set examples include 1/one, freedom-fighter/terrorist, Dad/Mr. President, Water/H2O, 1 divided by infinity/zero, energy/mass, two or more positions occupied by the same particle, and the world being probabilistic.

In the absence of compromise or coincidence growth or divergence of expressions must be caused by interest generation or a limitation or both. Similarly, any convergence of expression must be caused by interest consolidation or overcoming a limitation. In a world, with unified interests, there is no need for a labeling duality such as freedom-fighter/terrorist.

We can also propose that the road map towards truth might look somewhat like a pathway with the following stations/stops:

1. Detect and don't be fooled by coincidence,
2. Avoid compromises,
3. Remove limitations, and
4. Give up on interest.

We can also propose a progress metric. Conversion or diversion of expression and language may be a handy surrogate marker of progress towards or regression away from the truth. The unification of all spoken language might be an incredibly early indicator of us getting closer (a single universal language). Doing away with speech all together might be an advanced sign.

Later and Post-Model Afterthoughts:

1-   Einstein was on to something about the virtue of condensing language use when he said, "If you can't explain it simply, you don't understand it well enough". Even more to the point would be to paraphrase the quote to say, "If you have to explain it at all, you don't understand it well enough". This last quote is a hypothetical condition this writer predicts will happen when we get to unconditional certain knowledge.

2-   A hint of what lies ahead might be illustrated in the highly advanced language of Groot. Groot's advanced civilization impressively did away with excesses in speech to the tune of resting all expression in these three words that are always said in the following exact order I then AM then GROOT: "I am Groot"

3-   Related to this passage is the uniqueness paradox that you can read elsewhere.

4-   In logic, it is said that "BUT" is another way of saying "AND". This is yet another symptom of a language limitation in which more than one expression means the same thing.

112. Asking too much of philosophy and perhaps too little of idea's body language:

Could the body language of ideas tell us something insightful that might be additive to what philosophy tells us? We should disagree with Descartes, who tried his best to establish a methodology of certainty for philosophy. By doing that he was asking too much of philosophy. As he did not recognize or acknowledge that parts of philosophy may be closer to medicine than to mathematics. Therefore, the certainty of mathematics will be hard to replicate, but medical methodology might be worthy of exploring, where the medical scientific community often depend on less than certain metrics such as the p-value and metanalysis to weigh in on blurry ill-defined real-life issues.

Later and Post-Model Afterthought:

Upon further later reading, Giambattista Vico appears to have made similar but non-identical comments to what was written above in this segment and elsewhere in this book.

However, Vico arrives at his objection to rationalism through his principle that states that truth is verified through creation or invention and not, as per Descartes, through observation. Or put simply we can only understand what we can make, to know stuff, we have to be able to disassemble and assemble it. This can be applied to our history and mathematics according to Vico. For him, nature is not a human production, and therefore we can think about nature, but we could not understand it.

This writer agrees with Vico to question the Cartesian method's generalizability. But this writing arrives there thru different pathways and also this writing goes further and has a wider scope of objection that can have several forms:

1- This writing brings up that the Cartesian method (including the "cogito ergo sum") does not escape what all human knowledge (including the Verum Factum of Vico) have in common, conditions and condition boxed dynamics, that cannot escape internal self-referencing which in turn inevitably will cause scope limitations, anomalies, contradictions, paradoxes, or collectively a breakdown of some sort.

2- In the "Betweenness Model" of reality, another objection is suggested in that there is no making in reality. Things are either present or repurposed from something already present to start with.

3- In the "Betweenness Model" of reality, another objection is suggested in that the conserved elements of everything are completely incomprehensible and the interaction forces between conserved elements are partially incomprehensible.

113. Can physicists be trusted enough, to part ways with some well-established intuitions:

Quantum mechanics allows things to be in two or more places at the same time. It seems that this is fundamentally wrong, and it may turn out to be just a matter of one place being perceived as two loci due to an error of perception. Double vision does not mean that there are two of everything.

It is hard to pick sides anymore, it is harder to be impartial when you like one side more, and it is hardest when the side you like more is in error.

But hold on a minute, it is not hard after all, quantum mechanics allows for two or more positions at the same time. What a relief for it to be needless to choose sides, but instead pick them all, they are all right.

Later and Post-Model Afterthoughts: After later reflection, the author no longer holds some of these opinions.

114. Never is never right:

Barriers to truth include time and language. These barriers will always stand in the way and will "never" be breached, making truth eternally elusive.

115. The un-conserved necessary follow a synthesis and is ruled by change and vanishing

If we take all things that are not conserved, a necessary synthesis is followed by a necessary change, then we find that vanishing is an overwhelming dominant property compared to stay-ability or survival. For every X that stays there are many more Xs that have vanished. For every idea that stays there are many more ideas that have vanished. For every species that stays there are many more species that have vanished. For every living thing that stays there are many more living things that have vanished. For every you, at the moment there are many other yous in moments past that vanished. And so on.

For the conserved, it is all staying, while synthesis and vanishing have no place. One of the best comprehendible examples that can be mentioned is that for every measure of energy that stays there are no measures of energy that have vanished. (Momentum, charge, and angular momentum seem to be other approximations of the conserved). Incidentally, while synthesis and vanishing are each usually more than staying for the un-conserved, it is different for the conserved, where synthesis is less than staying but equal to vanishing in being nonexistent.

Between the conserved and the un-conserved exists a vague but fascinating relationship. It is difficult to put your finger on it, but this is an attempt at cracking this mystery:

1. For starts, it should be pointed out that there is an asymmetry in this relationship where it is an absolutely one-way relationship, in which the un-conserveds are necessarily dependent on the conserved to be, while the un-conserved has no influence whatsoever on the conserved that is completely sheltered from any un-conserved advances.

2. Furthermore, the conserved and the un-conserved can have a relationship where they have nothing in common or on the other hand they may have one where they are intertwined and difficult to distill from each other, but yet always have an element of strangeness and the feeling that something about this seemingly harmonic and seamless intertwined arrangement re-

mains off. It seems the former lack of common ground situation is what we should expect when the un-conserved lacks any extension outside of the mind (Using commonly used general descriptions or definitions: it can be said to be furthest from the truth and realness). While the latter intertwined situation is what we should expect when the un-conserved possesses or is nearing the truth and realness*. Additionally, in the latter, the un-conserved's nature or history is in some way generated, inspired, molded by the conserved, or at least used the conserved as a frame of reference, and this is where its truthfulness and realness stem from*.

3. It is difficult to plot this relationship to one particular association patterns. It is in some instances tight and at other times the un-conserved appears to be loosely anchored or underpinned to the conserved. The strength of this underpinning relationship is hard to measure but it seems that as the relationship becomes tighter, we might expect the un-conserved to be more conserved-like, truthful, and real*. As stated, it is difficult to gauge the strength of this relationship, but one metric is to look at the stayingness of things to associate it with their conserved-likeness, which may be used to infer truthfulness and realness.

4. Or look at things vanishingness to associated it with their conserved-unlikeness, which may be used to infer untruthfulness and unrealness.

5. This relationship does not appear to follow one pattern perfectly, suggesting that there is a factor or more that may play a role in stability and stay-ability other than the tightness of the underpinning or the strength

of un-conserved's anchoring to the conserved. It is suspected that some of this is due to the observer's abstract frame of reference (see below illustration).

6. Some relationship patterns that might be pictured, include an all or none underpinning relationship of the conserved with the un-conserved, or on the other hand, we can imagine that this relationship is in the form of a gradient of some sort.

It appears from the above, that a pattern emerges to suggest that the conserved is timeless since it just stays, is not synthesized, nor does it vanish. Notice that synthesis and vanishing are time-dependent qualities. Therefore, the conserved can be defined as what is timeless, has not been synthesized, nor has vanished. The un-conserved can be defined as what operates within time, has been synthesized, and may or may not have vanished. Synthesizability and vanish-ability are qualities of the un-conserved, while the conserved is completely void of them.

Everything we know is confounded by time which as this writer has pointed out is a quality reserved for the un-conserved, which is looking suspiciously unreal and disparately untrue as understood. It is no secret that this writing questions the realness of spacetime as understood, but it would be disingenuous if that is stated without some reservations. It seems that spacetime might be saved because it might be anchored, molded, or shown to be underpinned to the conserved in two ways:

1- Through "Noether's theorem", where time is suggested to be related to energy and space is suggested to be related to momentum. This would favor a juxta-truth or truth-adjacent designation to spacetime.

2- The "betweenness model" of reality. In this model, the

above spacetime and realness are perfectly consistent by considering spacetime a force that removes many of the objections to its realness.

116. Classification of understanding:

A. Unconditional understanding (Knowledge of the conserved or knowledge that is grounded in the conserved):

Almost certainly not obtainable.

What needs to be done to obtain unconditional understanding is unknown, but it might look something like this (also see below illustration): It starts with going in the opposite direction and increasing the conditions to gaining certainty at the expense of scope, in the hope that this certain or near certain knowledge might give us insight on how to overcome the limitations of our conditions. Increasing certainty in understanding or truth might start with the conditional understanding variety and then making it less prone to being un-factual by fact cross-checking with understandings that were governed by (near or) non-overlapping and (near or) un-related conditions. From there we can hope to build a solid foundation and gain new insight that may allow us to relax the conditions towards fewer conditions (a step closer to being unconditioned).

How to find understanding with completely or near completely un-similar conventions, is unknown, but it might require us to fact cross-check with a non-human non-human possibly an extra-dimensional created consciousness. This strategy seems to be one way to control for omnipresent confounders. Another hypothetical way for us to gain entry into this class of knowledge is for us to abandon our humanity and leave the restricting conditions that come with humanity behind with it.

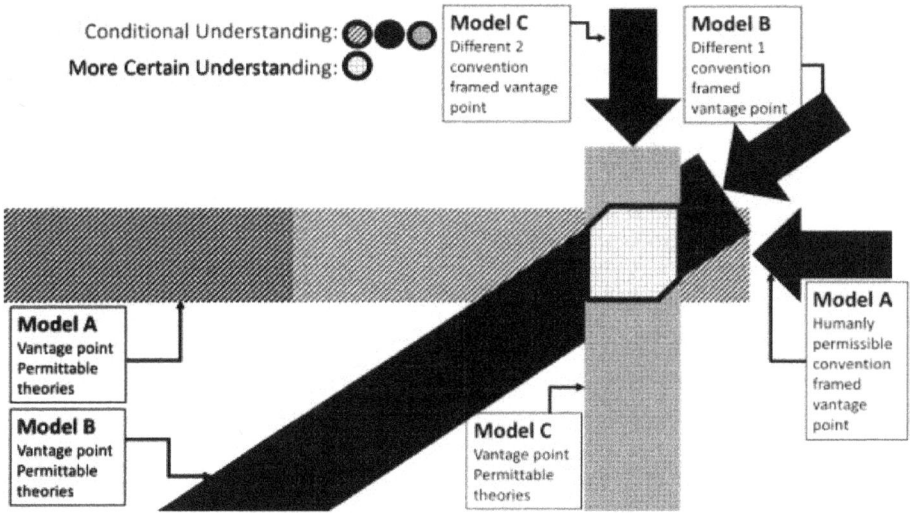

B. Conditional understanding (ambiguous knowledge that might not even be wrong): Is all meaning short of A.

1. Juxta-Truth (unconscious certified understanding): Unfathomable amounts of such facts already exist and continue to grow. These are obtained by us or our creations fact cross-checking with the "unconscious" without conflicting with A. Examples include some physics, meteorology, chemistry, etc.

2. Undifferentiated-Truth (the purely conscious understanding): Unfathomable amounts of such facts already exist and continue to grow. These are truths that are not and don't conflict with A or B1. Typically obtained by us or our creations fact cross-checking with human "consciousness" or human-created consciousness. Examples include logic, algebra, and mathematics.

3. Pseudo-Truth (unconscious rejected understanding): Unfathomable amounts of such facts already exist and continue to grow. These are ideas that conflict with A and B1. Examples include a perpetual motion machine, a forecast of raining unicorns, the philosopher's

stone, and whatever I say. (if this sounds familiar it maybe because it might be another way of expressing the "Ignorance Accelerated Inflation Model" we light-heartedly discussed earlier)

Conflict resolution simple guide:

	Juxta	Undifferentiated	Pseudo
Juxta	A least one is B3	A least one is B3	Within definition
Undifferentiated	A least one is B3	A least one is B3	B2 might be B1
Pseudo	Within definition	B2 might be B1	No change

In keeping with the spirit of the unconditional understanding the "betweenness model" of reality has dropped a condition in the definition of a force, so it follows that if the above structure of understanding is valid that the "betweenness model" might be closer to the truth by being less condition bounded.

117. Forecasting a minimum fact cross-checking requirement

This segment is speculation based on a couple of observations (A & B). It is also assumed that there is a time freeze, time correction, or time can be dismissed as a dimension.

A) How does our visual apparatus sample a 4-dimensional world? It accounts for time by taking multiple samples temporally. That leaves us with 3-dimensions. It receives input from 2 eyes, that accounts for one dimension. That leaves us with 2-dimensions. It gathers information on the retina's surface as 2-dimensions. That takes care of the rest. All the dimensions are accounted for approximately. Notice that the fact cross-checking is satisfied with two sets of information (from each eye).

So, it works out as: 4 dimensions of reality = 2-dimensional measurement (2D retina surface) + cross check factor of 1 (two eyes = 1 independent cross check) + controlling for 1 dimension (time). Or 4 = 2 + 1 + 1.

B) We can do the same thing with the technique our fellow radiologists use to locate things on x-rays. It takes two views (like the two eyes). A single Xray view is 2 dimensional. So, it also works out as 4 = 2 + 1 + 1.

It is almost certain that someone has worked this out more precisely but being too lazy to do the needed homework to find out, let us instead design a formula for ourselves using the above observations to know the minimum fact cross-checking requirement. Which can be stated as:

(Minimum number of fact cross-checking vantage points) =

(Reality number of dimensions) – (Number of dimensions of vantage point measurement) – (Number of controlled dimensions)

If our world was 3 dimensional, we should do as fine with one eye if it acquired two-dimensional data.

118. Effect of fact cross-checking

Two effects can be described when fact cross-checking is applied between two systems with non-identical conventions:

1- Error and bias cancelation. (Illustrated below by eliminated false theories and accurately cross-checked theories)
2- Addition of limitations. (Illustrated below by eliminated true theories and falsely verified theories)

It is always the hope that (1)'s gains outweigh (2)'s losses.

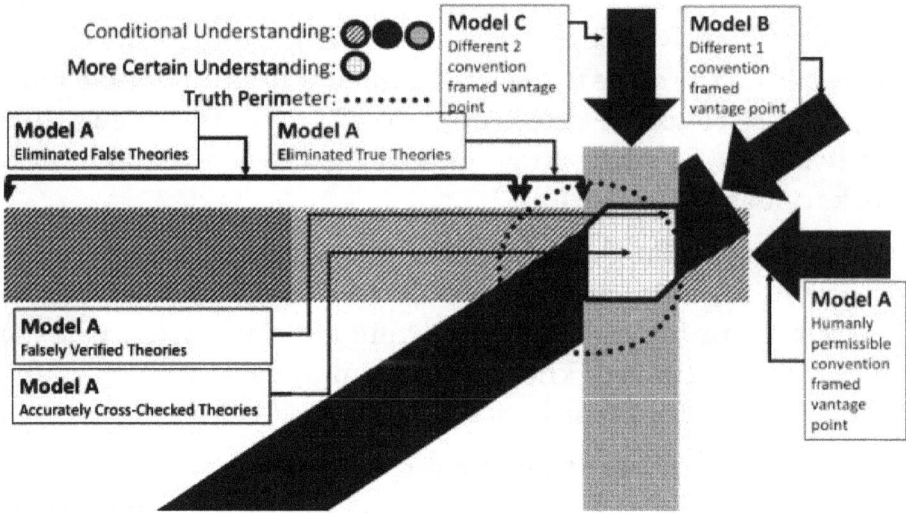

Conditional Understanding: ◐●◉
More Certain Understanding: ○
Truth Perimeter: ·········

Model C
Different 2 convention framed vantage point

Model B
Different 1 convention framed vantage point

Model A
Eliminated False Theories

Model A
Eliminated True Theories

Model A
Falsely Verified Theories

Model A
Accurately Cross-Checked Theories

Model A
Humanly permissible convention framed vantage point

119. Poor man's fact cross-checking

Not having non-humans (e.g. aliens) to cross fact check with is no excuse to not due with something else. Physicists have used several tools such as symmetry to cross fact check. This writing throughout the text has tried to correlate between several convention non-identical champions. These are physical reality, DNA, biology, brain, consciousness, ideas, language, convention, hypothetical artificial intelligence, and hypothetical artificial consciousness (See Table).

	Unsimilar Conventions and Presuppositions
Observable Reality	Time & Space
Biology	DNA/RNA Code
Brain	DNA/RNA Code + Ideas
Mental Language	Brain, People, and Consciousness
AI	Human mental conventions such as Code and Abstracts

120. Camouflaged metaphysics:

One example of language (or convention) restraining understanding is describing light and everything for that matter as being both particle and wave. Since our toolbox of descriptor tools had these two best-fit instruments that individually do a poor job but together do a better job. But the notion that Matter and Energy is intrinsically a mixture of both is unlikely and a contradiction. Matter and energy are things that our toolbox lacks the proper single descriptor for as well. Here, our language (or convention) distorts reality in our understanding.

Maybe an example will make what we are trying to say easier to understand. When a person with a squint sees you as two people, this does not mean that you are located at two places simultaneously but rather that the visual apparatus is dysfunctional, similarly, our physics and mathematics are dysfunctional as light is not both a wave and particle. Particle theories and wave equations have similarities to the sanitized imaginary Platonic abstract "forms" that attempt to explain reality. Their author and some followers have taken them to be literal and real and reality to be a derivative or shadow of the "forms". Waves and particles are similarly taken by many to be literally waves and particles.

Progress has not stopped, so there is the possibility that the world would come to find an upgrade to their toolbox (maybe the "betweenness model", fingers crossed) where the fate of the current tools (e.g. particle-wave descriptor tools) are retired in a similar manner to the retirement of the Platonic forms when something better came along.

Language/conventions in some sense are nothing more than well-camouflaged metaphysics (like the Platonic forms).

From this, it seems safe to predict that physics will progress in a more or less stepwise manner we can label as progress through "refinement" (which more or less parallels our investment in research and labs such as accelerators) up until this pattern is overshadowed and broken with a dramatic shift we can label as "disruptive" that is disproportionate to any investment. This may coincide with the advent of new tools or language (e.g. a mathematics successor, wave-free and particle-free theories, and equations).

121. Convention boxed dynamics

This might be a good point to go deeper and attempt to explain the preceding segment. The problems illustrated in the "camouflaged metaphysics" segment have developed because a convention boxed dynamic (e.g. science, language, thinking, code) does not fact cross-check with something outside its convention boxed dynamic. This closed system must indulge in self-referring no matter how well designed, even if the conditions are vastly inclusive, claimed to be self-evident, and self-correcting. It will inevitably lead to scope limitations, errors, contradictions, paradoxes, and anomalies. Some of these will be obvious but others will be invisible to the dynamic participants because these errors grew and evolve to satisfy the original conditions, evade detection, and not triggering any safeguard alarms, therefore understandably the participants in the dynamic might be blind to them.

This can be used as a counter to the most solid of arguments even the "Cogito, ergo sum".

The cross-checking requirement might be humanly impossible and a fatal human flaw to understanding but being impossible should not be an excuse to not acknowledge the lack of external fact cross-checking as a potential wrench into the validity of said convention boxed dynamic.

In illustrative terms, the "convention boxed dynamics" problem is a variant of the self-referring paradoxical statement "this sentence is false".

Remaining within the confines of the system conditions if feasible is often used as a practical strategy to carry on with any dynamic or system including life itself. However, sometimes you accidentally or forcefully venture beyond the convention limits then get faced by the unexpected and surprising. Or in other words, you are blindsided, entered a dead ally, or in keeping with this book's theme witnessed a "model breakdown". This can lead to a lot of possible results including fascination, confusion, struggles, unwelcomed insight, conflict, relief, excitement, or additional discoveries.

Maybe thinkers and philosophers (that have a somewhat incompletely overlapping convention boxed dynamic to science) should be having more than a spectator role to the unfolding of discovery and grow out of being intimidated and their submissiveness to not listen to each other. It might be time for them to call foul play on irrational and wild unchecked claims and trust that their set of un-similar conventions is a reason for them to have a limited but still independent vantage point which might serve a somewhat external useful fact cross-checking function to the otherwise closed self-supervised (science or philosophy) dynamic. The science community of course will say that science is "different", it is "not like the others", and it is "special". They will reason with science being self-correcting, which is true. But it is foolish to think that there is a perfect match between the "self-correction capabilities" with the "convention boxed dynamic error footprint". Some errors must slip through the self-correction safeguard.

The ultimate fact-checker should lack all convention. The second best would be to cross-check with a dynamic with non-overlapping conventions, which almost certainly will have to be non-human and non-human derived. The third-best would be a dynamic with partially overlapping conventions (e.g. logic and science), which might or might not be human. See the illustration in segment number: 118.

122. Legitimacy of questions

All questions must honor the condition limits of any system dynamic. Questions beyond any system dynamic conditions may be illegitimate. Where and when type questions are permitted, with an asterisk, within system dynamics that include or presuppose time-space, but in any system dynamics beyond time-space, those questions might be illegitimate. Since this writing suggests that time and space are forces (explained elsewhere) on the bases that they are not conserved entities, this leaves us to consider that we in part might occupy a spaceless and timeless system dynamics in which all where and when questions   don't   apply   always.

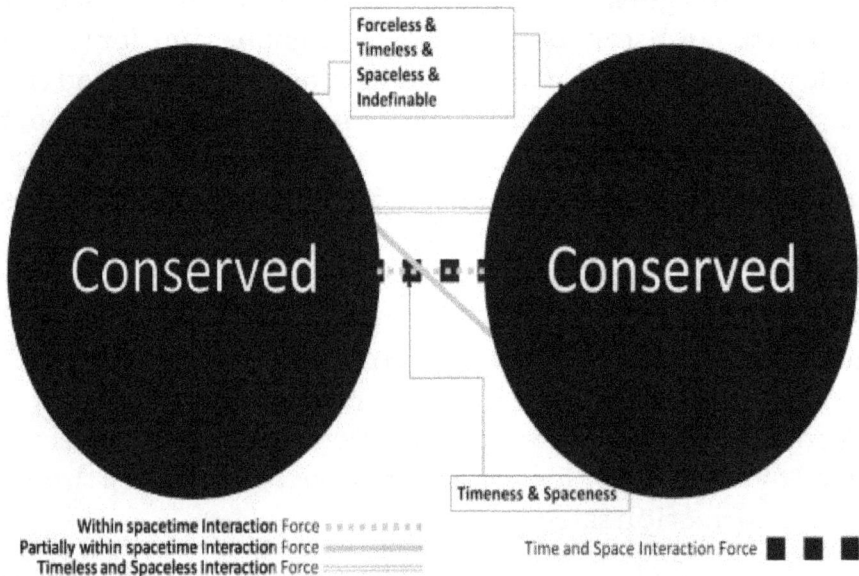

Forceless &
Timeless &
Spaceless &
Indefinable

Conserved   Conserved

Timeness & Spaceness

Within spacetime Interaction Force
Partially within spacetime Interaction Force
Timeless and Spaceless Interaction Force

Time and Space Interaction Force

123. When parents are
caught having sex

Every one of us has accepted something or another as being

angelical or free of imperfections by whatever standard. Those moments that you realize those highly held things are no longer so, can be monumental and are to the displeasure of most of us. They are irreversible and one-directional, where there is no way back for you to unsee, unhear or unknow what you just saw, heard, or knew. This writer was irreparably damaged when that happened to the most sacred of sacred, the most beloved of beloveds. It happened to no other than geometry, arithmetic, and logic.

It was not shocking to learn of efforts to validate non-Euclidean geometry by reducing it to Euclidean geometry. However, it was somewhat surprising that the validity of Euclidean geometry needed to be reduced to arithmetic to remove doubt. But things went too far, and it was nothing short of shocking to know of the need and efforts to validate arithmetic by reducing it to sets and logic.

The last part was pioneered by Frege who seemed to figure out a way to validate arithmetic, until his efforts were smashed at first only to then be salvaged by Bertrand Russell (who added a cast system where he conditioned logical statements to be divided into first-order statements, second-order, ... and prohibited mixing different classes or casts order statements as a "safeguard"). This heroic mind-bending breakthrough was only short-lived until it was too smashed by Kurt Gobel, who used a mirroring technique of different orders of statements to bypass Russell's "safeguards" and introduce contradictions, and the house of cards came down crumbling.

The simpleton in me looks at this and says to himself why does arithmetic need to be grounded? Why do logicians think that logic is an indubitable suitable anchor? Heartbroken doesn't even come close to describing what it felt like to see the fate of these once heroes of certainty fall from grace. (elsewhere we have demonstrated that "logic" too measurably breaks down too).

Idea & Gene Free Zone (Where you can't be)

You
(Where you are never left alone)

Ideas   **Genes**

Mental Divide

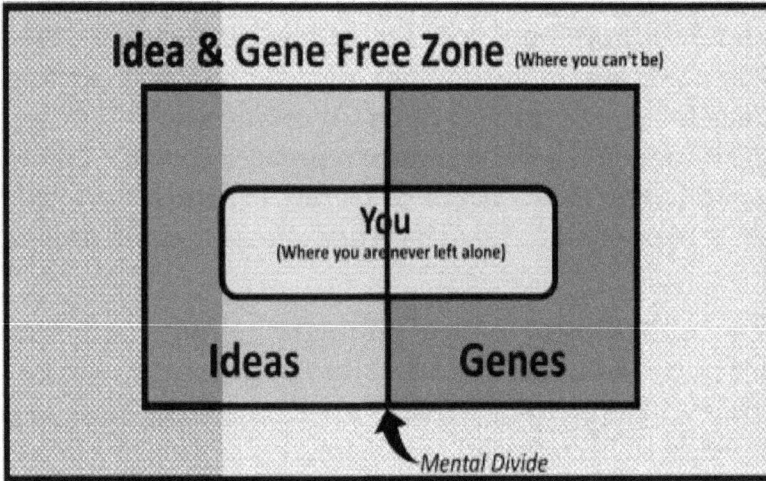

124. Meaning and ignorance
from where no one expected!

To the best of my knowledge, I do not know if no one expected it, but I know I did not expect to stumble on what you are about to read.

There is a focus in this writing on pointing out languages limitation and this might distract from noticing something that language does that it is not supposed to do. Language is so amazingly interesting and complex. This books opinion of language comes in two flavors:

1- Language is limited in representing meaning (Can mostly be found above ↑)

2- Language has the power to create or hide meaning (Can mostly be found below ↓)

Language is supposed to be a communication instrument that describes things; however, it is not supposed to create meaning where there isn't and frame, mold, or restrain understanding where meaning is free. Language although it is primarily a byproduct of meaning and understanding, "language" sometimes does the opposite where meaning and understanding are created as byproducts of "language" and where meaning, and understanding are destroyed due to "language".

Examples:

For an example of a language creating meaning and as mentioned elsewhere, look no further than the lines of stock price support and resistance at whole numbers and multiples of ten. These imaginary floors and ceilings have no relationship not whatsoever with anything fundamentally meaningful to the stock value. Also do this thought experiment for fun, without changing anything fundamentally real even a bit, these lines of resistance and support would be quite different with an abstract or language switch such as using Roman instead of Arabic numbers. We can also wonder about switches such as liter vs gallon, foot vs meter, Dollar vs Euro vs Yen, mile vs kilometer, lightyear vs kilometer, or stone vs pound vs kilogram.

As an example of language destruction of meaning, and as mentioned elsewhere, look at jokes that often lose their humor when delivered in a different language.

You can also try to describe Kareem Abdul Jabbar's height in centimeter vs lightyears (218 centimeters vs 2.3043e-16 lightyear) and see if one brings out meaning while the other one hides it.

You can also notice the meaning destruction of having something priced at 9.99 instead of 10.

General Description:

In physics, the "observer effect" is the theory that the mere observation of a phenomenon inevitably changes that phenomenon. This is often the result of instruments that, by necessity, alter the state of what they measure in some manner.

The Hawthorne effect also referred to as the "observer effect" is a type of reactivity in which individuals modify an aspect of their behavior in response to their awareness of them being observed. Both of these examples and other types of observer effects can undermine the integrity of research.

This writing in the above few paragraphs has described something not un-similar but instead of "observer" as an effector we have "language" and instead of "phenomenon" or "behavior" as the effected we have "meaning".

So, we can call this (to the best of this authors knowledge, never before described) theory "expression tampering or effect" or "language tampering or effect" that states that:

"The mere use of expressions or language to describe meaning inevitably changes that meaning"

In comparison to the other tampering effect, it is similarly omnipresent and inescapable. However, it probably is as if not more profound than the "observer effect" and "Hawthorne effect".

This line of thinking if projected further, can lead to the terrorizing and haunting thought that some aspects of physics such as spacetime with its units of meter and second might be a creation of "expression tampering" into meaning and understanding.

Explanations:

Two explanations are presented:

1- Is it not unreasonable to say that abstracts including language, mathematics, and geometry (tools that are mind-based) have created a certain framework or specifications that dictated how laws need to be for them to be the most compatible and non-contradictory to those specifications and framework they developed from and within.

Let us take a look at the visual apparatus and sight. Is it

a coincidence that our eyes and visual apparatus (with its conventions and languages that allow it to function) see the part of the electromagnetic spectrum that happens to match the most abundant wavelengths to come out of the sun? No, it is not. The spectrum of light that comes from the sun (you can think of expressions and conventions such as language in the same way) molded our visual apparatus development and functioned to frame its specifications. Then if we go one step further, we see that this visual and eye apparatus's internal methods of operations including its language and convention in by itself creates a framework that can and will mold the production of abstracts that are friendly to this convention and language framework. From this, it should come to no one's surprise that visually inspired abstracts such as artwork are more likely to be limited to the narrow electromagnetic wavelength framework (red, blue, yellow, ...), and seldom intentionally produced with paint or colors outside the visible light range (Do you remember seeing an artist intentionally using infra-red or ultraviolet paint to create her abstract art? I have not).

If paintings had little choice but to be made with colors within the visible range, maybe physics had little choice but to be made with situational restrictions too.

If a painter is unlikely to imagine an art piece with infra-red or ultra-violet, then maybe a scientist is unlikely to imagine a theorem or law with full range.

2-   The "betweenness model" or "conserved-force model" of reality identifies conventions and language as forces. The model gives us a platform for explaining the "language tampering effect" because the model identifies convention and language as forces that result in an action or reaction.

It is hard to visualize, but the illustration below attempts to bring it in focus better. The illustration crudely attempts to explain the above-mentioned stock price behavior, as it shows how we can imagine that lines of force might augment or cancel each other if they cross paths. It is believed that this and similar phenomena are the results of an interference effect that is well established in optics but the "betweenness model" of reality predicts is behind unfathomable poorly explained phenomena like the stock price behavior described above being an interference-effect. Expression tampering and language tampering are other suspected examples of interference phenomena or effects.

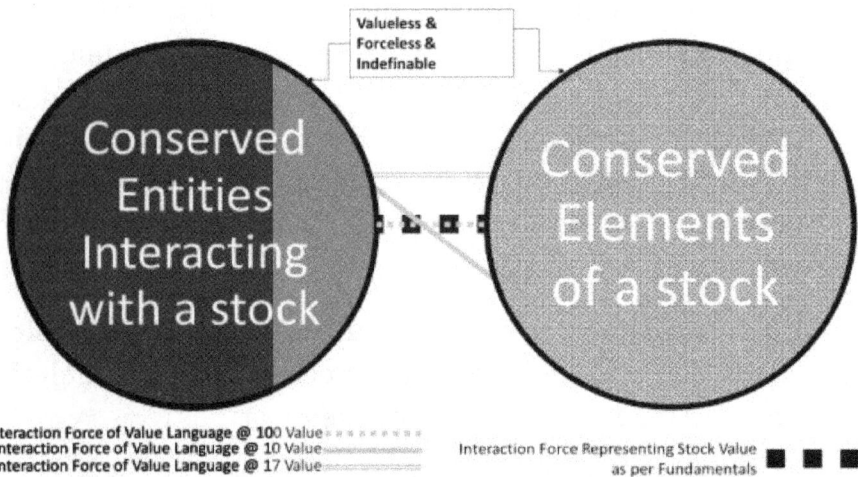

Similarly, the below illustration is a visual approximation of the "betweenness model" or "conserved-force model" of reality at work, where the above-mentioned joke varying funniness can be seen to be augmented or diminished depending on the type of language.

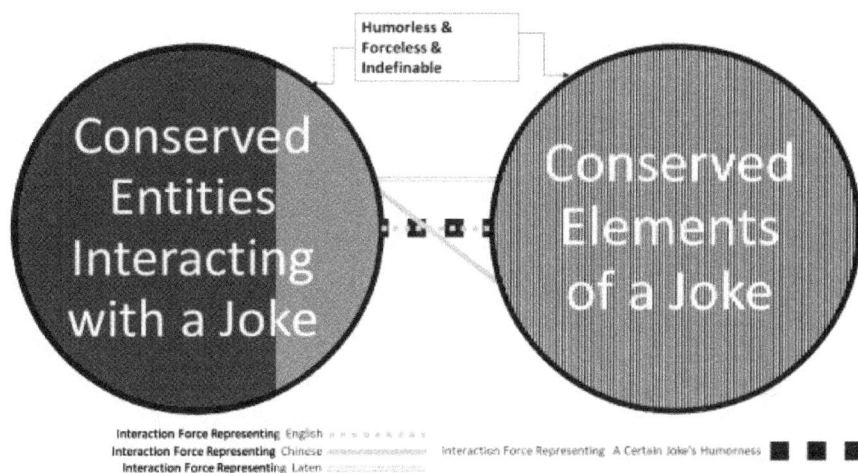

Interaction Force Representing English ○ ● □ ● ● ○ ● ○
Interaction Force Representing Chinese ～～～～～～～～～   Interaction Force Representing A Certain Joke's Humorness ■ ■ ■
Interaction Force Representing Laten ～～～～～～～～～

Suspected hidden and anticipated expression (or language) tampering:

Has something unsettling but similar occurred to other abstracts like arithmetic, mathematics, geometry, algorithms, set theory, laws of physics, relativity, quantum mechanics, string theory, and all others? The answer can only be yes.

What if the Big Bang is a physics version of a stock price level of resistance or support at rounded numbers such as 100?

Ironically, this book/writing that brought attention to language and expression tools tampering with understanding, maybe the reason that this writing (with its mathematically and linguistically unfriendly concepts) is unfairly treated compared to relativity and quantum physics (linguistically and mathematically friendly concepts) due to a "language tampering" effect that may have created, nurtured, molded or contributed to the creation of relativity and quantum physics to be the way they are (who can exclude a subtle, stealthy, and constant butterfly-like effect of "language tampering" in meaning and understanding that influenced or created things as important as relativity and quantum physics).

It then dawned on me not to forget that this influence can go the other way. Let us not forget what language might be hiding from us or which meanings and understandings may have been inhibited by language tampering.

One more added perk of acknowledging the language tampering effect is that it can be an awesome excuse for failures. This is how it may work: if humanity doesn't accept my hypothesis it might be due to an expression tampering effect as much as it is a bad idea. In other words, not my fault, it is a good idea, but people are manipulated to not appreciate it.

As a consequence of the expression tampering effect:

It exposes the need to add a new frame of reference to the other established frame of references. In physics a frame of reference is used, often embodied in an abstract coordinate system. In Einsteinian relativity, an observer is used as a frame of reference. What if the collective or fundamentally important human abstracts restrict us to a limited vantage point or frame of reference that makes many things like relativity, quantum physics, and spacetime appear realistic. These same theories might not be realistic if viewed with new sets of eyes (see vantage points or models A, B, & C in the below illustrations). We can call this concept the "abstraction frame of reference" that in some ways is like Einsteinian "observational frame of reference" or other physics "frame of references". Also see the figure in segment number: 118.

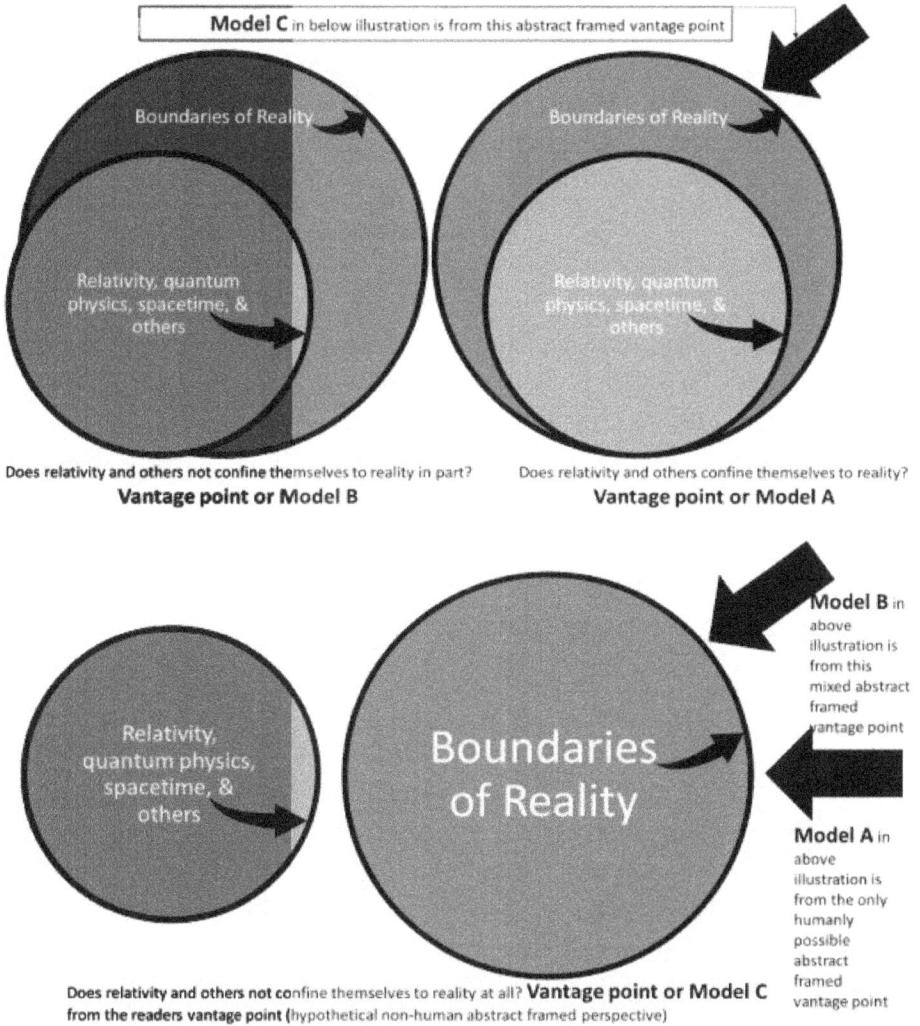

Model C in below illustration is from this abstract framed vantage point

Boundaries of Reality

Relativity, quantum physics, spacetime, & others

Does relativity and others not confine themselves to reality in part?
Vantage point or Model B

Boundaries of Reality

Relativity, quantum physics, spacetime, & others

Does relativity and others confine themselves to reality?
Vantage point or Model A

Relativity, quantum physics, spacetime, & others

Boundaries of Reality

Model B in above illustration is from this mixed abstract framed vantage point

Model A in above illustration is from the only humanly possible abstract framed vantage point

Does relativity and others not confine themselves to reality at all? **Vantage point or Model C** from the readers vantage point (hypothetical non-human abstract framed perspective)

We can take this further and project humanly inaccessible theories, and we can see that our theories may be fabrications of vantage point, convention, or language. You can say an ultimate example of the "language or convention tampering effect" described above.

What if:

We made a hunting sport out of findings examples of convention, expression, or language tampering into meaning. What if it grows in popularity? Then what if that leads to competitions and events or even championships "Language Tampering Hunting Championship". It would be interesting to see what strategies champions might employ, but it might be based on interference mining (See heading number: 56).

125. Omnipresent confounders

There are things everywhere and everywhen that we do not consider (not that it is possible) controlling for in our attempts at understanding. Not being able to control for things is one thing but operating as if controlling is an important instrument of discovery then allow exemptions without mention, disclosure, or acknowledgment can be misleading and is probably the root cause of accepting uncertainty as the way nature is. These include time, space, mind, consciousness, genes, language, ideas, conditions, and the un-conserved.

126. Uniqueness paradox

Firstly, since quantum physics taught us that things can be in more places than one at the same time with a non-zero probability. Who knows maybe all things, events, places, times co-exist always where they are all-wheres and all-times in a state of "allness"? Think of all words in a book being there since the beginning but realizing them is a human-generated (a quality that the words and book are innocent of) temporal illusion because of the way we read a book not because the book did not contain the words in their entirety (not within human's capacity) from the onset.

Secondly, we are told (or at least I imagine) spacetime coordinates are unique. If I say the original Mona Lisa will be at 10 longitudes, 10 latitudes, rounded sea level to the higher meter increment, at 1/1/2222 noon GMT, it would be a unique spacetime coordinate. The mail is based on address uniqueness where a letter to be sent to France will not cause surprise if it reached

France, but would cause a reaction if it ended up in Japan, but would be insanely confusing if it was delivered to both countries at the same time.

These two, presuppositions are contradictory and hence the uniqueness paradox. One thing must give in. This puts us in a forced position where we can question spacetime uniqueness and realness, question matter and energy existence, or question our sanity and understanding including quantum physics and logic.

Another way to think about it is what was emphasized in this book, and that is that the spacetime position/coordinate is not conserved and therefore there is doubt about its realness as defined now. From this, we can say that the uniqueness paradox pertains to an unrealistically defined entity (spacetime) and is, therefore, not a paradox at all. Also, related to this segment is a segment titled "which one is real? Neither".

127. Simulation twist ending (spoiler: it's us, not aliens, demons, or computers)

You hear of these convoluted often conspiratorial simulation theories by others such as supercomputers, aliens, a demon, etc... In these theories, we experience a fake simulation while believing it is real.

This author thinks we can stop the speculation about simulation because it is factual and undeniable.

Furthermore, we should just remove the non-us (others) actors from the discussion. Adding others is redundant and unnecessary in keeping with the spirit and not violate Occam's razor (Occam' razor: More things should not be used than are necessary). Why bring in computers, aliens, and demons to accuse of tricking us thru a simulation, when all we have to do is simply point a finger at ourselves as the perpetrators and victims at the same time as a simulation.

The factual claim comes from the realization of indisputable human self-simulation instruments we refer to as abstracts (including language, geometry, mathematics, laws). Abstracts are so pervasive in our existence that it is not possible to distinguish real from a simulation. All abstracts are simulations that add a filter to our experience. We should stop speculating about simulation anymore, as it unquestionably can be removed from the conspiracy theory box to take its proper place among the facts of life.

Conspiracy theory box nomination: Before leaving this topic, it would be an oversight if this book does not nominate a never before considered idea to the conspiracy theory box. It is perhaps the ultimate conspiracy namely "Self-Trust".

128. Irreversibility of some knowledge

In some states of mind, you can believe the pointlessness of it all, then you see parts being fabricated and assembled into a purposeful AK-47. Then you see a subset of knowledge seemingly always growing and being refined as if there is a single directionality in its materialization, where some information appears resistant to reverting or reversing direction back to the unknown, suggesting it might be under the influence of an underlying law that dictates one-directional unfolding of some knowledge subset, not un-similar to the second law of thermodynamics where entropy always increases. Learning of an AK-47 may not be willfully unlearned, could be an example of one-directional knowledge. This puts to ease the pointlessness tantrum I started this paragraph with, sorry for that.

129. A root cause of all progress

It is knowledge being defined as:

"justified true belief"

We are not saying here that the definition is the cause of progress, but rather the other way around, in that the definition embodies and is one manifestation of what has and continues to govern human progress.

The definition may be a subtle example of the Texas sharpshooter fallacy (shoot your firearm at a barn then draw a target sign around where the projectile hits). The definition was constructed to fit and be friendly to what we have achieved. The definition was not purely constructed objectively while blinded to human achievements. The definition proceeded with significant human achievement and progress; hence the Texas sharpshooter fallacy is suspected. You can say the definition is a confession of what all our human progress is built on. A definition that is likely rooted in rationalization, justification, and fulfillment of some sort of wishful thinking that framed this definition.

By switching the word justified with certain, the definition becomes:

"certain true belief"

Here, we have wiped out all knowledge, which becomes an impossibly attainable definition of a "certain true belief". It is hard to envision anyone claiming let alone celebrating progress that was achieved without knowledge.

We essentially looked at what we did and are doing and molded the definition to not take away our knowledge, instead of showing discipline and being impartial by telling it as it is (For a nomination or example of an objective definition: see the part we talked about "unconditional knowledge" discussed elsewhere).

How we describe knowledge will not change much no matter

how hard we bang on the table, because the definition makes us feel better at the expense of accuracy, a price we have shown time and time again that we are more than willing to give up.

130. We can't be trusted.

People profess throughout the ages that they value knowledge greatly. It is hard to take that notion at face value nor should we dismiss it too. It is extremely hard to not believe that it is a genuine desire.

Without questioning intentions, it is interesting that other things appear to have progressed more or less the same or have progressed sometimes much greater than knowledge. Two causes emerge. The first is that progress is not dependent on professed desire alone. The second is that professed desire is inaccurate.

More to the second element, looking at how far we collectively have progressed might be impressive at first, but it seems less than expected given our professed love of knowledge and learning. For example, consumption, over-eating, drinking, smoking, pollution, overpopulation, entertainment, the sexual revolution, or conflict are candidates that may have outpaced knowledge.

Let us not forget about the recent globally tremendously successful adoption of the world wide web and social network platforms. This undoubtedly has increased the volume of available resources of information and increased overall human and personal exchange of ideas by many folds. After decades of such a multifold increase in usage. Was there a matching improvement in our collective knowledge? The answer is no. It might even be fair to say that there is no clear winner between information and misinformation.

This leads us to the question of why is a multifold jump in resources and communication not mirrored with a matching dividend of knowledge? What stands in their way to be more impactful concerning learning? The tremendous rise of the internet and digital social instruments without a matching rise in knowledge strongly means that we are more interested in the act of exchanging and interaction than actual substantive wisdom yielding exchange and interaction, otherwise, we would have collectively have become much wiser. We can't be trusted; we are just acting as if wisdom matters.

131. Enslavement as measured by language

An independent does not need language. The more language you use the more you signal to us how enslaved you are. Anything expressed with language testifies to your enslavement to the listener. The author is enslaved by caring hence the wordy book.

132. A law in jeopardy

In unpublished work, a proposed law states "products cannot understand their creator". The closest thing to break this proposed law is our attempt at understanding DNA and RNA. Here there is a chance that we may understand one of our creators. This sparks a few comments.

Why is this law at risk? It seems that this law is betting on our inability to get close enough to unquestionable unconditional knowledge. So, it stands to reason to think that the law of "products cannot understand their creator " seems potentially shaky because there is a real opportunity to get close to unquestionable unconditional knowledge with the study of DNA and RNA.

What could be behind us getting closer? Not sure, but based on the philosophy of this book, there is an emphasis on cross-checking to gain confidence in knowledge. Furthermore, there are qualities of cross-checking, where the best is what involves the least overlapping conditions and presuppositions. The pursuit of DNA and RNA understanding seems to be cross-checking two languages with potentially sufficient non-overlapping conditions and presuppositions (e.g. our consciousness on one hand and DNAs unconsciousness on the other hand). Those two languages with different "condition boxed dynamics", which are the "mental language" of understanding, science, and human knowledge, and the "non-mental language" of DNA and RNA.

Is there potential more significance to this? What we are witnessing in the study of DNA and RNA might be our best effort yet at getting close to true unconditional knowledge. It might provide a methodology, roadmap, and blueprint to the remaining scientific and wisdom seeking causes on the value of cross-checking with something outside of itself or something with a different "condition boxed dynamic" with less overlapping conditions, which provides a situation that may take advantage of system weakness and vulnerability cancellation.

What does the "betweenness model" or "conserved-force model" of reality have to say about this? From this perspective, there should be no risk to the proposed law. There should be no risk of the creator being understood. The creator is a combination of conserved (non-in-betweens) and non-mental forces (in-betweens that have extensions that go beyond the brain). On the other hand, understanding is purely a mental force (an in-between that has no extension beyond the brain). From this, it can be seen that the jurisdiction of understanding cannot reach all aspects of the creator preventing it from the opportunity to fully grasp what made it.

133. Consciousness black holes

Existing consciously is a convention and condition boxed dynamic that will cause inconsistencies and paradoxes necessarily. These will lead to dead ends that for all practical purposes are consciousness inescapable black holes.

It is an eternal trap unless the conscious being has the tools, power, and escapes velocity needed to break out of the hold of the black hole. Escape tools may come in many forms such as, memory loss, delusions, irrationality, coping mechanisms, impure reason, setting consciousness off-limits, a subconscious, interpretation errors, noble lies, and complaining. Otherwise, consciousness will be unstable and brittle. This is the fate that this writing predicts for AI-generated consciousness if it materialized.

Upon further reading after writing the above paragraph independently, there appears to be some expert support to this segment's proposed "consciousness black hole" and the elsewhere described "condition boxed dynamic" problems.

For instance, there is the "halting problem" which describes the problem of a computer entering an infinite loop forever. This cannot be predicted in advance. It was shown by Alan Turing that it is impossible to provide a general solution to all possibilities using algorithms. It sounds like computers can and in a large enough sample will walk in a dysfunctional state unknowingly. If this sounds familiar you are not wrong. The halting problem and infinite loops sound like special case examples of the "condition boxed dynamic problem" and the "consciousness black hole problem".

Admittedly, we cannot know if consciousness is immune or has found a solution to the mathematically proven halting or infinite loop problems. But it might be useful to humbly describe how humans perhaps have overcome the "halting problem" in three ways:

1- By being not all mathematical or algorithmic. The human solution appears to have taken the non-mathematic non-algorithmic form of memory loss, delusions, irrationality, impure reason, setting consciousness off-limits, a subconscious, interpretation errors, noble lies, and complaining.

2- Elsewhere, we also talked about a built-in halting (death and memory loss, unavailability, or suspension) and in-exact resets (offspring conception and relearning or memory recovery) that humans possess to make them possibly less susceptible to the "halting problem".

3-   The "betweenness model" or "conserved-force model" of reality paints biological consciousness's method of operation in a manner where it has little in common with a computer's method of operation. This makes the halting and infinite loop problems seem uniquely non-biological problems.

134. Prophecies will turnout foolish, more often than not

It appears arrogant predicting the eventuality of an AI take over based on the assumption that human brains are number-crunching computers with a ceiling that will be surpassed in due time. Several concepts need to be unpacked. Taking over has two main forms:

1-   It could be an unconscious runaway process that got out of control and caused human harm or extinction. This is possible. In essence, it is not different than unconscious biological or nuclear weapons causing human harm or extinction. Or unconscious greenhouse gases causing a runaway greenhouse effect that makes earth uninhabitable.

2-   Secondly, it could be in the form of a conscientiously deliberate AI take over from humans. In this case, the jury is still out on its possibility. For this to happen AI must be self-aware and become conscious first, which no one can know is possible or not. One reason that it is impossible to know, is that consciousness is not well understood nor does it have a satisfactory definition so how do you begin to build it or recognize it if it were ever to become. But this does not stop people from thinking it is an eventuality. Let us look at that a little closer. There are assumptions involved:

a. For example, firstly, it is assumed that consciousness is a mathematical, numerical, algorithmic, or a binary code product. This is not a given and is speculative. (Upon later reading, expert support was discovered. Roger Penrose deduced that some aspects of our brains are non-algorithmic, and the laws of physics are inadequate to explain consciousness. While computers are algorithmically deterministic systems)

b. Secondly, it is assumed that consciousness is purely a mind product (Model A in the below illustration). Part of the author sometimes wonders (could our consciousness be more in line with Model B in the below illustration) if there is an oversight of how relevant DNA and RNA are to the production of consciousness with the mind (perhaps thru a language tampering effect, described elsewhere). Or if the autonomic or peripheral nervous systems networks might be players too.

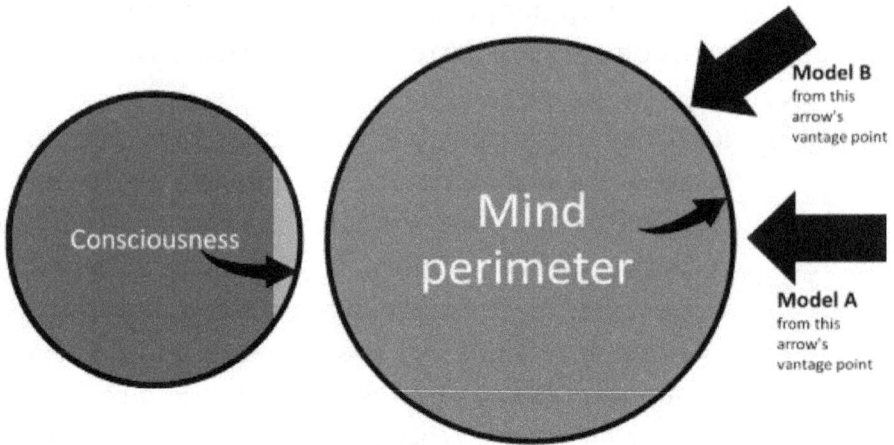

Does consciousness not overlap with the mind? **Model C** from the readers vantage point

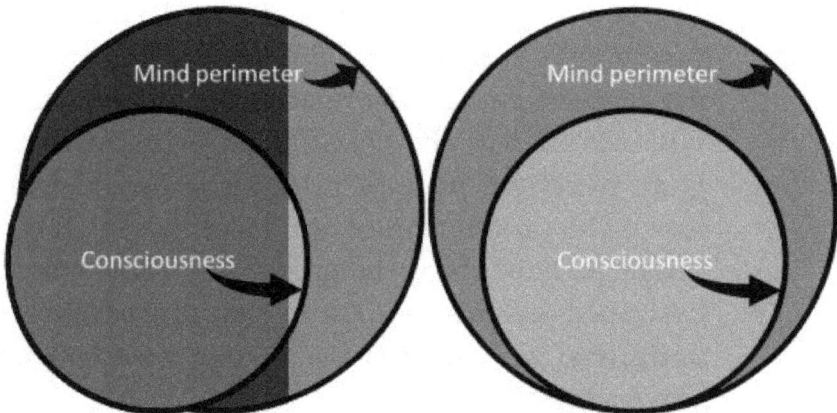

Does consciousness go beyond the mind? **Model B**

Is the consciousness limited to the mind? **Model A**

c. Thirdly, AI believers assume that consciousness is a product of a single language while it might be the product of two or more languages working together. What if the language of DNA and RNA may have something to do with consciousness? Then we must apply (2a) to it and assume that not only is the mind mathematic but that the language of DNA and RNA is also mathematical, numerical, algorithmic, or binary encoded. To the best of this writer's knowledge DNA and RNA is as mathematical as English, French, and Chinese are. In this instance, AI may not be as well-positioned or equipped to gain inroads into understanding or grasping the meaning of the words and sentences. This is difficult to explain but this story might make it a little more comprehensible: The original Spaniard invaders of America took the actions of those natives following them with incents as an indication of the visitors being viewed as divine and godly by the natives. Later, it became known that they were doing this because the natives did not know how to tell the Spaniard that they stink and needed to clean up so they used the incents to mask the awful smell of the visitor, of course, this comes down to a language barrier. Could consciousness not be mathematical causing the Spaniards amongst us to assume divinity, destiny, or a number-crunching eventuality. Don't worry if this story seems forced because it is, but it was not edited out because it is funny.

d. Fourthly, it is assumed that mathematics is the language of reality and truth. This is not a forgone conclusion. We need to remind ourselves that mathematics is still a language. Therefore, suffers from all the limitations of language that this book is littered with. This includes that it is not conserved, it is not timeless, it is a mind-made abstract and therefore of checkered trustworthiness. Therefore, unlikely to be able to grasp truth and reality nor consciousness which is a product of a non-mental reality. On a related topic this and similar segments in the book appear to construct a reasonable objection to the simulation arguments by Nick Bostrom.

e. Fifthly, some might say that AI might follow its road to consciousness without following the footsteps that the mind, DNA, and RNA took. This is fair and can bypass the need to understand the mind, DNA, and RNA. We still have some other problems:

    i. It does not absolve or excuse us from the need to understand consciousness or be able to define and recognize it if it ever materialized as mentioned in (2).

    ii. Most of us assume we are a product of reality or all the above discussion is meaningless if you start with the assumption that you are unreal and maybe a simulation. So, being a product of reality means that our consciousness is a product of reality and is anchored or has an underpinning in reality. If we assume for the sake of the discussion that AI acquired a consciousness, this AI consciousness is a product of AI, which is a product of computing, which is a product of a programming language, which is more or less mathematics, which is an abstract mind product, which means that AI consciousness is not non-mental and is anchored or has its underpinning in a mental abstract idea. If consciousness is not purely non-mental then AI can never attain it. We and AI do not share the same creator so temper your expectations.

f. Sixthly, it is assumed that consciousness comes in one flavor. What if the AI permittable flavors of consciousness are prone to be missing ingredients necessary for a "will to survive"? It seems that AI is not well-positioned to have some of these ingredients. These ingredients include:

   i. Progress might require a capacity to complain

   ii. Tolerating life and consciousness itself might require delusions and irrationality

   iii. Avoiding a paralysis state might require the capacity to set consciousness boundaries and off-limits

   iv. Poor interpretation of inputs might be required for the sense of purposefulness

These and other limitations and flaws on the surface are possibly important ingredients for survival under the surface. These can be inferred from such works as "the critique of pure reason" by Immanuel Kant, "the republic" with its inclusion of the concept of "noble lies" by Socrates and Plato, and Sigmund Freud's concept of the "subconscious". These may be necessary because existing consciously is a convention and condition boxed dynamic will cause inconsistencies and paradoxes necessarily. These will lead to dead ends that are for all practical purposes are consciousness inescapable black holes unless the conscious being has the power to escape in the form of delusions, irrationality, impure reason, setting consciousness off-limits, a subconscious, interpretation errors, noble lies, and complaining. AI consciousness if it materializes will likely be unstable and brittle unless it has the tools to escape the "consciousness black holes".

g. This is not a concrete absolute argument against AI consciousness, but AI consciousness requires some faith in something to happen that has never happened before an inconsistency generated consciousness (see below illustration).

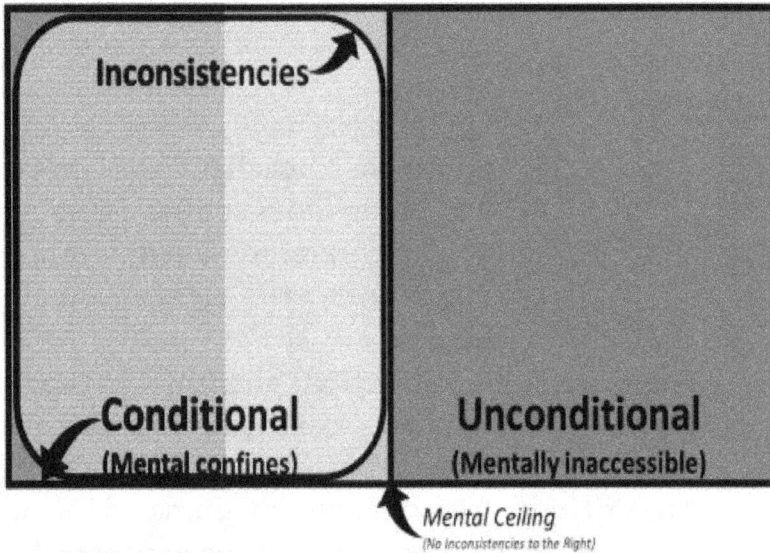

Mental Ceiling
(No inconsistencies to the Right)

If AI does overrun us it will likely be due to a mindless (maybe indistinguishable from mindful) runaway out of control cascade of events. The chances of a deliberate mindful dethroning are hard to imagine. If AI does take over it is more likely because humans willed AI to do so. It will be humans displacing themselves through a proxy agent.

135. Identity's indefinability

(Disclaimer: This segment's concepts are not original ideas, they are compiled from several sources)

Traditionally everyone is assigned one identity and is called a person or individual. This is looking less likely and challengeable.

In cognitive dissonance, there is the ability of people to harbor contradictory ideas and beliefs. Isn't this a feature you would associate more with two or a group of people?

In split-brain conditions, it has been documented that a person can compartmentalize knowledge where part of that person can know something while other parts are ignorant of it. Also, in the same condition, behavior can be telling in the sense that parts of a person can be seen to do something while the other part is indifferent or does something conflicting. Aren't these features something you would expect more with two or a group of people?

So, is it reasonable to see the traditional view of a person and identity being less likely and challengeable? Could a *person* be composed of *persons* sharing a body? Could an *individual* be truly a *dividual*? Where do we go from here regarding identity? It seems that each one of us is mostly a united front of a collective. Although understandably impractical, it becomes increasingly inaccurate to collectively label *dividual* persons who share a body and increasingly accurate to recognize the potential diversity of *dividual* persons who share a body. With this in mind you and me trying to define ourselves, especially with one-liners, now looks more laughable. People labeling others is even more so.

What would happen if you and I were ever free of our bodies? Would the only thing keeping these *dividual* persons together is a body? Would they want to be together? Do all these *dividual* persons have the same satisfaction or dissatisfaction? Are they fair to each other?

The above thoughts are consistent with the "betweenness

model" or "conserve-force model" of reality. The model recognizes the existence of a subset of forces that are brain confined and therefore mental. The model fundamentally defines a force as an in-between entity that connects non-in-betweens. Here we can see that there is a plurality of connected parts acting and reacting to the forces applied to them. A single identity does not sit well with this proposed model world view.

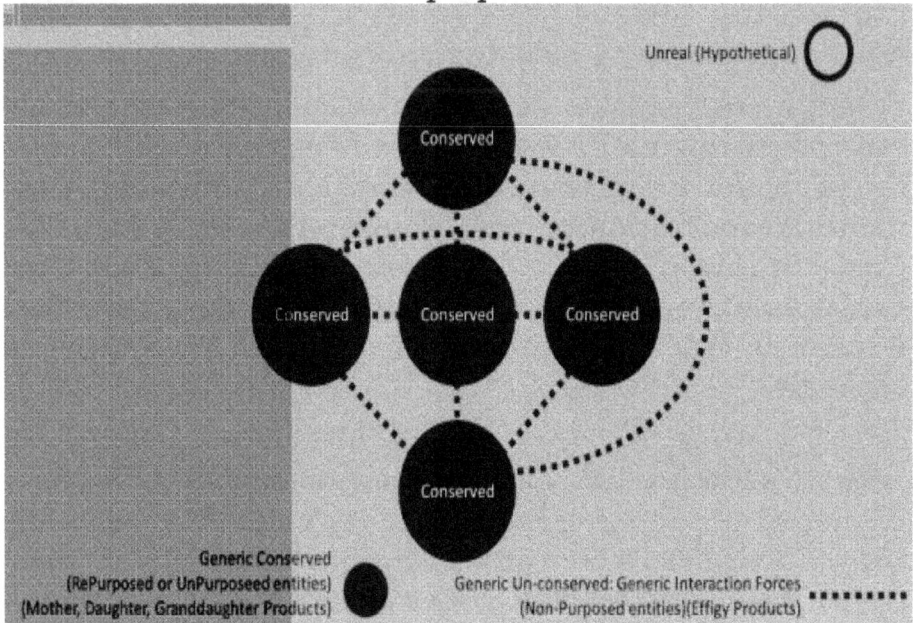

136. Convention Prison Break

All products are bound to fundamental identity-defining conventions. Escape from these heavily guarded convention walls is possible through three prison break strategies:

1. Death or suicide: Destruction or analysis (stop being)
2. Release: Losing Identity and seizing to be a product as originally defined (become something else)
3. Organ donor: Being repurposed to produce something with a different identity (become scrap parts to something else)

4. Escape: Identity preserving escape doesn't seem possible

137. Are conventions inherited from producer to product?

On an initial look, it seems so to some extent but not by necessity. A product is only obligated to keep its identity-defining conventions or risk losing its identity and becomes a product no more. Beyond that, the product is only limited by reality barriers. Therefore, a product can have fewer restrictions than the producer. This fits well with some observations we can make about the world:

1- For instance, DNA (producer) has never and will never be self-aware, yet it aided in producing an awareness (product). A human (producer) needs oxygen, yet a computer (product) does not.

2- Nothing is truly created but rather repurposed, therefore, the new products are just a rearrangement of elemental stuff that was there before. That elemental stuff is eternal, but the rearrangement and reorganization need not be.

In confession, the writer started thinking about this question hoping to demonstrate that a product is obligated by its producers' boundaries. Because that would be a further reassuring argument in the AI consciousness debate. But it seems as if this effort failed unless someone much capable has or will later show that such a limit existed in a way that escaped this author.

Practically, we should keep the AI consciousness debate within the confines of what defines AI, and not bring in its producers (brain) limitations in the discussion as if they apply in absolute permanent terms to AI. But this does not absolve AI from its foundationally defining conventions,

which necessarily include human mental languages that its
production is based on (e.g. binary, code ......).

	Human obligations	AI obligations Advantage goes to	
DNA conventions to say	Yes	No	Hard
Code conventions to say	No	Yes	Hard
Energy, matter, reality	Yes	Yes	Wash

www.ingramcontent.com/pod-product-compliance
Lightning Source LLC
LaVergne TN
LVHW011157080426
835508LV00007B/459